8th International Workshop on Information Optics 2009

AA008599

(WIO 09)

Journal of Physics: Conference Series Volume 206

Paris, France
20-24 July 2009

ISBN: 978-1-61738-271-0

Printed from e-media with permission by:

Curran Associates, Inc.
57 Morehouse Lane
Red Hook, NY 12571

Some format issues inherent in the e-media version may also appear in this print version.

Copyright© (2009) by the Institute of Physics
All rights reserved.

Printed by Curran Associates, Inc. (2010)

For permission requests, please contact the Institute of Physics
at the address below.

Institute of Physics
Dirac House, Temple Back
Bristol BS1 6BE UK

Phone: 44 1 17 929 7481
Fax: 44 1 17 920 0979

techtracking@iop.org

Additional copies of this publication are available from:

Curran Associates, Inc.
57 Morehouse Lane
Red Hook, NY 12571 USA
Phone: 845-758-0400
Fax: 845-758-2634
Email: curran@proceedings.com
Web: www.proceedings.com

TABLE OF CONTENTS

PAPERS

Partial polarization of light beams in time and frequency domains

Tero Setälä[1], **Ferdinando Nunziata**[2], **and Ari T. Friberg**[1,3,4]

[1] Helsinki University of Technology (TKK), Department of Applied Physics, P.O. Box 3500, FI-02015 TKK, Finland

[2] Università di Napoli Parthenope, Dipartimento per le Tecnologie, Centro Direzionale, isola C4, I-80143 Napoli, Italy

[3] University of Joensuu, Department of Physics, P.O. Box 111, FI-80101 Joensuu, Finland

[4] Royal Institute of Technology (KTH), Department of Microelectronics and Applied Physics, Electrum 229, SE-164 40 Kista, Sweden

E-mail: `ari.friberg@tkk.fi`

Abstract. The time-domain degree of polarization of a stationary, random optical beam differs in general from its spectral degree of polarization. We elucidate the origin of the differences and consider examples in which the two degrees of polarization either are, or are not, the same.

1. Introduction

The temporal coherence properties of a fluctuating, polychromatic optical field generally differ from its coherence properties evaluated at a certain frequency. For scalar fields relationships between the complex degrees of coherence in the space-time and space-frequency domains have been derived, implying that the two quantities are connected but have, as a rule, different values [1]. Further, a polychromatic optical field considered at a single frequency is often confused with a monochromatic field which by definition is fully deterministic, and therefore also fully coherent and polarized. A random polychromatic field at a certain frequency need not be either fully polarized or spatially fully coherent. In this paper we consider the distinctions between the partial polarizations of beam-like electromagnetic fields in the time and frequency domains [2,3]. We explicitly point out the origin of these differences and illustrate them by several examples.

2. Degrees of polarization in time and frequency domains

The time-domain polarization properties of a random beam field propagating in the z direction and represented by a zero-mean complex analytic signal $\mathbf{E}(t) = [E_x(t), E_y(t)]$ are contained in the equal-time coherence (or polarization) matrix of the form (Ref. [4], Sec. 6.2)

$$\mathbf{J} = \begin{pmatrix} J_{xx} & J_{xy} \\ J_{yx} & J_{yy} \end{pmatrix}, \tag{1}$$

where $J_{ij} = \langle E_i^*(t)E_j(t)\rangle$, $(i,j) = (x,y)$, with the angle brackets and asterisk denoting averaging and complex conjugation, respectively. The degree of polarization that characterizes the state of partial polarization of the beam is obtained from the equal-time coherence matrix, Eq. (1), as (Ref. [4], Sec. 6.3)

$$P_{\mathrm{t}} = \sqrt{1 - \frac{4\,\det\mathbf{J}}{\mathrm{tr}^2\mathbf{J}}}, \tag{2}$$

where the subscript t refers to the time domain.

The equal-time coherence matrix does not contain any knowledge of the temporal coherence or spectral properties of the field. For that information, and to eventually assess the spectral polarization properties, we make use of the electric mutual coherence matrix defined as

$$\boldsymbol{\mathcal{E}}(\tau)\begin{pmatrix} \mathcal{E}_{xx}(\tau) & \mathcal{E}_{xy}(\tau) \\ \mathcal{E}_{yx}(\tau) & \mathcal{E}_{yy}(\tau) \end{pmatrix}, \tag{3}$$

with $\mathcal{E}_{ij}(\tau) = \langle E_i^*(t)E_j(t+\tau)\rangle$, $(i,j) = (x,y)$, and satisfying various conditions [4].

The frequency-domain polarization properties are included in the spectral coherence matrix, obtained from $\boldsymbol{\mathcal{E}}(\tau)$ of Eq. (3) by employing the generalized Wiener-Khintchine theorem,

$$\boldsymbol{\Phi}(\omega) = \frac{1}{2\pi}\int_{-\infty}^{\infty} \boldsymbol{\mathcal{E}}(\tau)e^{i\omega\tau}\mathrm{d}\tau. \tag{4}$$

Making use of $\boldsymbol{\Phi}(\omega)$, we can introduce the spectral degree of polarization as (Ref. [5], Sec. 9.2)

$$P_{\mathrm{f}}(\omega) = \sqrt{1 - \frac{4\det\boldsymbol{\Phi}(\omega)}{\mathrm{tr}^2\boldsymbol{\Phi}(\omega)}}, \tag{5}$$

where the subscript f emphasizes that $P_{\mathrm{f}}(\omega)$ is defined in the frequency domain.

On inverting Eq. (4), setting $\tau = 0$, and noting that $\boldsymbol{\Phi}(\omega) = 0$ for $\omega < 0$, we find that

$$\mathbf{J} = \int_0^{\infty} \boldsymbol{\Phi}(\omega)\mathrm{d}\omega. \tag{6}$$

This equation, with Eqs. (2) and (5), shows at once that the time-domain degree of polarization P_{t} and the spectral degree of polarization $P_{\mathrm{f}}(\omega)$ have, in general, different values. However, if the field is fully polarized in time, it is also fully polarized at each frequency, i.e., $P_{\mathrm{t}} = P_{\mathrm{f}}(\omega) = 1$, whereas a temporally unpolarized beam may exhibit any spectral degrees of polarization [2].

3. Example

As an illustrative example, we consider a beam which is a superposition of two fields: a beam linearly polarized in the x direction and its delayed replica polarized in the y direction. Such a field can be expressed as $E_x(t) = A(t)$ and $E_y(t) = A(t - \tau_d)$. For the time-domain degree of polarization we find $P_{\mathrm{t}} = |\gamma_A(\tau_d)|$, where $\gamma_A(\tau) = \mathcal{E}_A(\tau)/\mathcal{E}_A(0)$ is the degree of coherence of the random process $A(t)$. We observe at once that when the time delay τ_d is much longer than the coherence time of $A(t)$, the beam is fully unpolarized, $P_{\mathrm{t}} = 0$. On the other hand, when $\tau_d = 0$, the field is fully polarized, $P_{\mathrm{t}} = 1$. Hence, by adjusting the time delay τ_d we can construct a beam whose time-domain degree of polarization is anything within $0 \le P_{\mathrm{t}} \le 1$. However, for any value of P_{t}, the field is fully polarized at all frequencies, as is seen by first employing the Wiener-Khintchine theorem and then Eq. (5) to find that $P_{\mathrm{f}}(\omega) = 1$ for all ω.

References

[1] Friberg A T and Wolf E 1995 *Opt. Lett.* **20** 623
[2] Setälä T, Nunziata F and Friberg A T 2009 *Opt. Lett.* **34** 2924
[3] Lahiri M 2009 *Opt. Lett.* **34** 2936
[4] Mandel L and Wolf E 1999 *Optical Coherence and Quantum Optics* (Cambridge: Cambridge U. Press)
[5] Wolf E 2007 *Introduction to the Theory of Coherence and Polarization of Light* (Cambridge: Cambridge U. Press)

A degree of freedom and metric approach for non-singular Mueller matrices characterizing passive systems

Vincent Devlaminck and Patrick Terrier

LAGIS, UMR CNRS 8146, Université Lille 1, Sciences et Technologies, 59655 Villeneuve d'Ascq, France

e-mail: vincent.devlaminck@univ-lille1.fr

1. Introduction

The description of Mueller matrix set is always questionable, thought a part of the solution is already well known: non-singular non-depolarizing Mueller matrices are the positive multiples of Lorentz group matrices with positive left upper corner element and positive determinant (according to the used definition of Mueller matrix). Then, corresponding sub-group is given [1][2] by $\{SO(3,1)_e\} \cup \{G\,.SO(3,1)_e\}$ with G = diag(1,-1,-1,-1). Therefore, the characterization of matrices of $SO(3,1)_e$ (the identity component of $SO(3,1)$, the associated special orthogonal group) completely describes the two sub-groups. Extension of this result to the whole set of non-singular Mueller matrices is not straightforward since this set will be expected to form only a semigroup [3]. In a recent paper [2], regarding mathematical framework to define and analyze a general parametric form of an arbitrary non-singular Mueller matrix, we proposed to address this problem in a space with more than 4 dimensions in order to introduce a set of transformations with a group structure. We consider a group of transformations continuously connected to the identity transformation and preserving a N-vectors norm such as the proper Lorentz transformations with the 4-vectors of the Minkowski space.

2. Appropriate norm and dimension.

The problems with such approach are then: how do we determine N, the space dimension and how do we chose the appropriate norm?

2.1. Appropriate norm

The proposed space should contain a Minkowskian subspace associated with the four first components for instance. Then, this subspace can be assimilated with the polarization space - it means $(x^0, x^1, x^2, x^3) = (S_0\ S_1\ S_2\ S_3)$ - and the associated metric should be defined by the metric tensor $g_{\mu\nu}$ with the signature $(1,-1,-1,-1, g_{44}, g_{55}, \ldots, g_{(N-1)(N-1)})$. Considerations about properties of the polarizance vector or diattenuation, allow us [2] to demonstrate $g_{44} = g_{55} = \ldots = g_{(N-1)(N-1)} = -1$. So, we are now considering the group $O(N-1,1)$ and more precisely $SO(N-1,1)_e$, its identity component.

2.2. Appropriate dimension

What is under interest is the number of degrees of freedom (DOF) associated with the polarization 4-dimensions subspace not with the full N-dimensions space. As $SO(N-1,1)_e$ is a matrix Lie group with

N degrees of freedom, we can focus our attention on the N generators of the group rather than the infinite number of group elements. We prove that N=7 is the actual solution. Under this hypothesis, we have now a potential of 21 DOF. So how many DOF are actually present in the resulting Mueller matrix? We first divide these generators in 4 sets and consider the DOF contribution to Mueller matrix subspace of each set and the DOF added by the interaction between these sets under the matrix multiplication rule. The first set is composed by J_{23}, J_{13}, J_{12}, J_{01}, J_{02} and J_{03}. The 6 generators give associated Mueller sub-matrices with a total of 6 DOF. In fact these Mueller matrices are the non-singular, non depolarizing Mueller matrices (product of M_R, the retardance matrix and M_D the diattenuation one [4][2]), since these generators are translation in 7 dimensions of Lorentz group generators. The same approach with J_{46}, J_{56}, J_{45} shows that no DOF can be directly added to the Mueller sub-space or by the block matrix multiplication rule either. We now consider the J_3 matrix defined by linear combination of 9 generators J_{m4}, J_{m5}, J_{m6} with m =1,2,3. This set of generators gives an amount of only 6 DOF. The last set of 3 generators gives the matrix J_4 as a linear combination of generators J_{06}, J_{05} and J_{04}. It is straightforward to demonstrate no DOF can be directly added to the Mueller sub-space by these generators. But it is clear that DOF can be added to the Mueller sub-space by the block matrix multiplication rule between $\exp(J_3)$ and $\exp(J_4)$. These matrix products add a diattenuation vector or a polarizance vector [4] with 3 DOF to the Mueller matrix subspace. Eventually, we obtain a total amount of 15 DOF for the resulting Mueller matrix subspace. The same approach proves we only have an amount of 14 DOF for the Mueller matrix subspace when dealing with a 6-dimensions space. Then $SO(6,1)_e$ is a physical admissible solution to solve the question of definition of arbitrary non-singular Mueller matrices. The previously proposed solution in [2], was just a subset of this one and it is possible [6] to identify missing matrices as ones with equal depolarization factor along the 3 principal axis with a non null polarizance or diattenuation vector. It is worth noticing that the relation between the both definitions of Mueller matrices and the different connected components of $O(3,1)$ introduced for the non-depolarizing matrices is not so straightforward for the case of depolarizing matrices since $SO(6,1)_e$ is a group of matrices with unit determinant but sub-matrices of this group can have negative determinant.

3. Conclusions

Starting from previous results about the non-depolarizing Mueller matrices, we generalized the method to any non-singular Mueller matrices. Considering the set of non-singular Mueller matrices as a set of linear operators on an embedded subspace of a more general space, we take metric and degree of freedom in consideration in order to specify this space and the group of associated linear transformations. Generators of this group are used to address the issue of the dimension of the full space and a 7 dimensions Minkowski space gives a physical admissible solution. Arbitrary non-singular Mueller matrices are then sub-matrices of $SO(6,1)_e$

References

[1] D.H. Sattiger & O.L. Weaver, Lie Group and Algebras with Applications to Physics, Geometry and Mechanics, (Springer Verlag, Inc., 1991)

[2] V. Devlaminck and P. Terrier, "Definition of a parametric form of nonsingular Mueller matrices", J. Opt. Soc. Am. A 25, 2636-2643 (2008)

[3] R. Sridhar and R. Simon, "Normal form for Mueller matrices in polarization optics", J. Mod. Opt. 41, 1903-1915 (1994)

[4] A. A. Sagle, R. E. Walde, Introduction to Lie groups and Lie algebras, (Academic Press, 1973)

[5] S. Y. Lu, R. A. Chipman, "Interpretation of Mueller matrices based on polar decomposition", J. Opt. Soc. Am. A 13, 1106-1113 (1996)

[6] V. Devlaminck, P. Terrier, Non-singular Mueller matrices characterizing passive systems, Opt. Int. J. Light Electron.Opt. (2009), doi:10.1016/j.ijleo.2009.06.001

Journal of Physics: Conference Series **206** (2010) 012003 doi:10.1088/1742-6596/206/1/012003

Liquid Crystal Light Valves for Slow Light and applications

S. Residori[1], U. Bortolozzo[1] and J.P. Huignard[2]

[1]INLN, CNRS, Univ. de Nice Sophia-Antipolis, 1361 route des Lucioles, 06560 Valbonne, France

[2]Thales Research & Technology, RD 128 91767, Palaiseau Cedex, France

E-mail: jean-pierre.huignard@thalesgroup.com

Abstract. The large dispersive properties of wave mixing in liquid crystal light-valves allow obtaining fast and slow light with tunable group velocities. A slow light interferometer is shown by using this interaction.

1. Introduction

Slow and fast light have recently received a large interest, both from the fundamental point of view and for applications in, communications, high sensitivity interferometers, precision metrology and optical sensing [1]. Different methods to realize slow and fast light are currently employed. Among them, the processes of nonlinear wave mixing offer the advantage of large and tunable dispersion properties, which can be used to obtain controllable group velocities in small-sized experiments. These properties have recently been exploited in photorefractive crystals [2] and in optical fibers through stimulated Brillouin scattering [3]. Here, we show that by performing two-wave mixing experiments in a liquid crystal light valve (LCLV), we are able to obtain fast and slow light down to group velocities as slow as a few tenths of mm/s. In the LCLV the two-wave mixing occurs in the Raman-Nath regime of diffraction, leading to multiple-order output beams, each characterized by a different group delay [4].

2. Two-wave mixing in the liquid crystal light valve

The liquid crystal light-valve combines a thin liquid crystal (LC) layer (thickness 10 μm) with a monocrystalline Bi12SiO20 (BSO) photoconductive crystal (1 mm thickness, 20x30 mm2 transverse size).The other wall is a glass window as shown on fig 1. Transparent electrodes allow the application of an external voltage across the liquid crystal layer.

The photoconductive and electro optic properties of the LCLV are separately optimized. While the excellent sensitivity comes from the photoconductive properties of the BSO, the large electro-optic effect is due to the birefringence of the liquid crystals. The BSO is transparent in the visible range and, thus, the LCLV works in transmission. This property allows performing wave-mixing experiments, in which two or more beams interact inside the LC layer and gives rise to nonlinear phenomena such as optical amplification and phase conjugation.

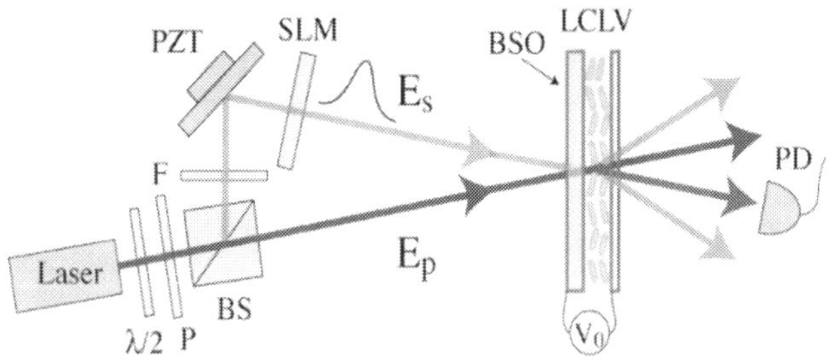

Figure 1. Schematic representation of the two-wave mixing occurring in the liquid crystal light valve.

Two-wave mixing in the LCLV occurs following the schematic representation of Fig.1. A pump beam Ep and signal beam Es are sent to the BSO side of the LCLV with parallel polarizations along axis of the LC molecules to create an interference pattern. The beam coupling occurs in the Raman-Nath regime of diffraction and several diffracted beams are observed at the output of the LCLV. The intensity of the pump beam is much higher than that of the signal. For this condition, part of the pump is transfered to the diffracted orders: the output signal on the m=0 order results amplified, with a gain G >1 and there is a phase transfer on the signal as a function of the pump-signal frequency detuning leading to slow light interaction.

3. Slow and fast light in the LCLV: tuning the group velocity

To obtain slow and fast light the 2WM is performed by sending on the LCLV a light pulse together with the pump beam. The pulse width is larger than the LC response time, which is around 120 ms and its center frequency can be changed by a few Hz by using a piezo mirror. By exploiting the large phase dispersion we can obtain both fast and slow light, with the output pulse either anticipated or delayed. In Fig.2a) and b) are shown two representative datas, displaying a fast pulse taken at the m=0 output and a slow-light output at the m=-2 diffraction orders. The dispersion curves can be theoretically calculated from the model of the two wave mixing in the LCLV [4].

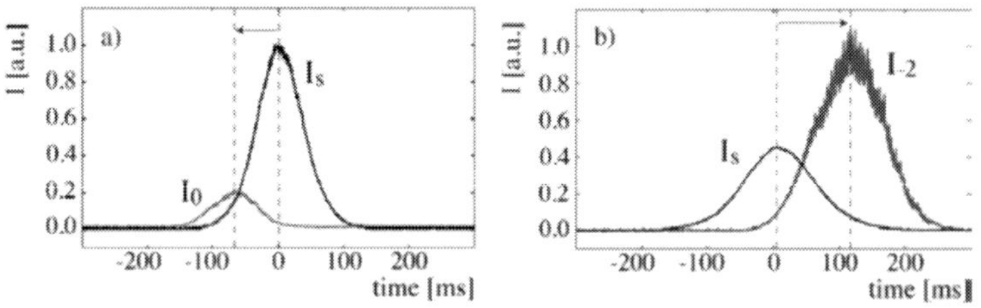

Figure 2: Experimental time dependencies of the output pulse for the m=0 (red) and m=-2 (blue) diffraction order; the output pulse is anticipated and, respectively, delayed, of a) 65 ms, b) 110 ms.

4. Interferometry with the LCLV as a slow light medium

The spectral sensitivity, or resolution, can be enhanced by a factor equal to the group index ng of the slow light medium [5]. At the same time with LCLV, the optical amplification prevents losses and provides a gain that is tunable with the applied voltage, so that the slow light interferometer can be adjusted to work in optimal conditions. As example, we have realized a Mach-Zehnder interferometer, where a LCLV in inserted in one of the arms. We use the m=0 order, so that the signal is present even when the LCLV is not operating and we can make a direct comparison between the sensitivity with and without the slow light effect.

5. Conclusions

In conclusion, we have shown that by performing two-wave mixing experiments in LCLV, and by using the dispersive properties associated to the two-beam coupling process, we can obtain fast and slow light phenomena with very large and tunable group delay. The corresponding group index due to slow light has also been exploited to enhance the sensitivity of interferometric detection systems.

References

[1] R.W. Boyd, and D.J. Gauthier, "Slow and fast light", in Progress in Optics, edited by E. Wolf (Elsevier Science, Amsterdam, 2002), Vol. 43, pp. 497-530.
[2] A. Shumelyuk, K. Shcherbin, S. Odoulov, B. Sturman, E. Podivilov, and K. Buse, "Slowing Down of Light in Photorefractive Crystalswith Beam Intensity Coupling Reduced to Zero", Phys. Rev. Lett. 93, 243604-243607 (2004).
[3] M. González-Herráez, K.-Y. Song, and L. Thévenaz, "Optically controlled slow and fast light in optical fibers using stimulated Brillouin scattering", Appl. Phys. Lett. 87, 081113 (2005).
[4] U. Bortolozzo, S. Residori and J.P. Huignard, "Self-pumped pumped phase conjugation in a liquid crystal light-valve with tilted feedback mirror ", Opt. Lett. 32, 1-3 (2007).
[5] Z. Shi, R.W. Boyd, D.J. Gauthier, and C.C. Dudley, "Enhancing the spectral sensitivity of interferometers using slow-light media", Opt. Lett. 32, 915 (2007); Z. Shi, and R.W. Boyd, "Slow-Light Fourier Transform Interferometer", Phys. Rev. Lett. 99, 240801-240804 (2007).

Nanophotonics for information systems

Y. Fainman, K. Ikeda, M Abashin, D. Tan

Department of Electrical and computer Engineering, University of California, San Diego, 9500 Gilman Drive, La Jolla, California 92093-0407

Email: fainman@ece.ucsd.edu

Abstract. The field of photonics finds applications in information technology, health care, lighting, and sensing. This paper explores the role of nanotechnology with focus on nanophotonics in dielectric and inhomogeneous metamaterials.

1. Introduction

Optics has the potential to solve some of the most exciting problems in computing hardware. It promises crosstalk-free interconnects with essentially unlimited bandwidth, long-distance data transmission without skew and without power- and time-consuming regeneration, miniaturization, parallelism, and efficient implementation of important algorithms such as Fourier transforms. Numerous information processing systems and concepts in space and time have been studied during the past decades. Yet, optical computing and processing in space and time has so far failed to move out of the lab. The free-space and guided-wave devices are costly, bulky, and fragile in their alignment. They are also difficult to integrate with electronic systems, both in terms of the fabrication process and in terms of delivery and retrieval of the massive volumes of data the optical elements can process. Our most recent work emphasizes the construction of optical subsystems directly on-chip, with the same lithographic tools as the surrounding electronics. This has been made possible by the advances in these tools, which can now create features significantly smaller than the optical wavelength; experts predict lithographic resolution as fine as 16nm by year 2020. Arranged in a regular pattern, subwavelength features act as a metamaterial whose optical properties are controlled by the density and geometry of the pattern and its constituents. Lenses, polarizers, chromatic dispersers, diffraction gratings, and other optical processing devices can now be implemented on-chip using metamaterials wherever natural materials with similar properties either do not exist, or (more frequently) would not be compatible with lithographic fabrication.

2. Nanophotonics process

To advance this technology, investigations of nanostructures and their interaction with electromagnetic field are critical. Engineers also need appropriate modeling and design tools, new fabrication recipes, and test instruments capable of characterizing on-chip components. The design of integrated photonic systems is a challenging task as it not only involves the accurate solution of electromagnetic equations, but also the need to incorporate the material and quantum physics equations to enable the investigation and analysis of near field interactions. These studies need to be integrated with device fabrication and characterization to verify device concepts and optimize device designs. In this talk, we discuss some of the CMOS-compatible Silicon-on-insulator metamaterials and devices recently demonstrated in our lab. These include graded-index lenses, birefringent elements that utilize a combination of geometry and material properties to separate light into orthogonal polarizations, frequency-selective resonators and Bragg gratings, and metal-dielectric nanostructures that can achieve extremely tight field confinement. Characterization tools, including our Heterodyne Near Field Optical Microscope, will also be discussed. This microscope uses a fiber probe tapered to 200nm diameter and brought close enough to the nanostructure under test to pick up its evanescent electromagnetic fields. Subsequent heterodyne detection permits simultaneous measurement of both amplitude and phase of the evanescent fields, while also providing an amplification to boost the weak signal. The mapping of evanescent fields has proven to be a powerful aid in understanding the performance of nano-optical elements.

3. Example: Cladding modulation

Bragg structures are fundamental in implementing various optical devices for switching, wavelength division multiplexing (WDM) and sensing applications. We have previously demonstrated on-chip sidewall modulated Bragg grating (SMBG) realized using single-step lithography. The strength of the coupling coefficient in SMBG is determined by the modulation amplitude which in turn is limited by the resolution of electron beams used to pattern these structures. Narrow bandwidth WDM components require Bragg gratings which have weak coupling coefficients and which can be easily controlled. To address these issues, we introduce a novel cladding-modulated Bragg grating (CMBG) implemented with silicon on insulator (SOI) material platform. The proposed CMBG is shown to overcome the limitations inherent to SMBG in fabricating devices that require high resolution, wide dynamic range and high precision control of the coupling strength.

Our CMBG consists of a single mode waveguide at 1.55um with the Bragg effect arising from placements of silicon cylinders with period, Λ_B a distance, d away from the silicon waveguide (see Fig. 1). The cylinder radius, R is chosen to be 100nm to avoid supporting any resonant modes. Since the field amplitude of the propagating mode decays exponentially outside the waveguide boundaries, the extent of the evanescent tails residing in the silicon cylinders and hence the strength of mode coupling can be varied by adjusting the distance d. We first calculate the CMBG coupling coefficient, κ of the CMBG as a function of d using coupled mode theory (CMT).

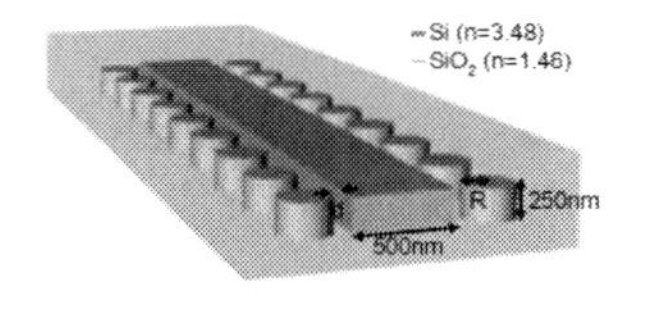

Figure 1. CMBG schematic. Waveguide radius, R=100nm

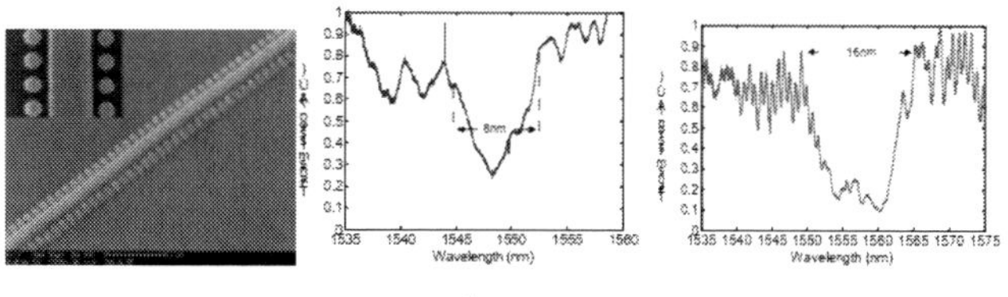

a b c

Figure 2. (a) SEM Micrographs of fabricated devices. Measured transmission spectra of (b) device A (L=100μm, d=200nm) and (c) device B (L=70μm, d=50nm)

The device is fabricated using electron-beam lithography followed by reactive ion etching and plasma enhanced chemical vapor deposition of the SiO_2 overcladding. The spectral response of two fabricated devices (A & B) measured using an optical spectrum analyzer is shown in Fig. 2. Even though the length of the device B was L=70μm, the value of $\kappa.L >> \pi$ indicating that κ is the dominant in contributing to $\Delta\lambda$. The values of $\Delta\lambda$ measured for devices A and B are 8nm and 16nm respectively, close to the expected values of 7nm and 13nm respectively from our 2D FDTD results.

Journal of Physics: Conference Series **206** (2010) 012004 doi:10.1088/1742-6596/206/1/012004

4. Conclusions

It is evident that nanophotonics and metamaterials are important technologies for future information processing systems. The research exploiting near field optical phenomena will be key in achieving new functionalities in materials properties (birefringence, dispersion, nonlinearities) as well as unique device functionalities such as adaptation, generation, modulation, transport and detection of light.
This work was supported by the NSF, DARPA, and the NSF CIAN ERC.

Role of spatial correlation in photon-counting integral imaging

SR Narravula[1], MM Hayat[1], and B Javidi[2]
[1] Center for High Technology Materials and Department of Electrical & Computer Engineering, The University of New Mexico, Albuquerque, NM 87131-1356
[2]Department of Electrical & Computer Engineering, The University of Connecticut Storrs, CT 06269-2157

E-mail: hayat@ece.unm.edu

Abstract. The method of photon-counting integral imaging (PCII) has been introduced recently for three-dimensional object sensing, visualization, recognition and classification of scenes under photon-starved conditions. This paper presents an information-theoretic model for the PCII method, thereby providing a rigorous foundation for our understanding of its demonstrated success in compressive imaging and classification.

1. Introduction

It is intriguing that 3D photon counting imagery captures much of the information or content in an image even under photon-starved conditions, as demonstrated by Yeom *et al.* This feature has become particularly evident and useful in the context of image classification, where very good performance is observed even when very few photons per pixel are available, Yeom *et al.*

If we consider the process of imaging by means of a photon-counting array, which we refer to as the method of photon-counting integral imaging (PCII) system, we can identify three main components in the system, the true intensity image or the "source," the transformation rule or the "channel" and the "output" photon-counting array, which counts the photons impinging on each detector element. We observe that the mutual information between the source and the output is a measure of the change in the information present in an image prior to transmission and that present at the destination [2, 3]. In this paper we will use the normalized mutual information metric, the mutual information between the source and output normalized by the source's entropy, applied to elemental images in the PCII process to investigate the role of spatial correlation in how well image-content is preserved.

2. Problem Formulation

Consider a stochastic column vector $\mathbf{X}$ whose entries, X_i, $i = 1, \ldots, n$, are discrete random variables in the interval $[0, 1]$ representing the reflectance of some unknown object or an unknown digital image. The number of photons detected per pixel in a unit integration time, Y_i is a Poisson random variable with mean value $x_i \epsilon N_p$, where $\epsilon \in [0, 1]$ is an attenuation factor that we will use as a control parameter in this study and N_p is the mean number of photons per pixel and per unit integration time.

Under photon starved conditions, the attenuation factor, ϵ, is very small and the photon counts per pixel are either 0 or 1 with high probability. The probability mass function of the photon count Y_i, conditional on the pixel's gray level $X_i = x_i$, takes the simpler form of [5]

$$P_{Y_i|X_i}(j|x_i) = (1-j)e^{-N_p x_i \epsilon} + j(1 - e^{-N_p x_i \epsilon}), \quad j \in \{0, 1\}. \tag{1}$$

3. Results

In all of our calculations we have used Markov models for the source image; these are are very good probabilistic models for capturing spatial correlation. We have also adopted the hard-limited model (1) and formulated the normalized information metric, ρ, analytically.

In the one-dimensional case, correlation is introduced in terms of a parameter, m, such that as m increases the spatial correlation increases between neighboring pixels. Figure 1(a) shows ρ as a function of the transmission probability, ϵ, parameterized by different spatial correlation indices, m. It is seen that performance metric, ρ, increases with spatial correlation in the source image. This suggests that the ability of the PCII approach to retain spatial information improves with the spatial correlation in the source image. Namely, the PCII approach seems to be inherently geared toward "images" rather than individual pixels.

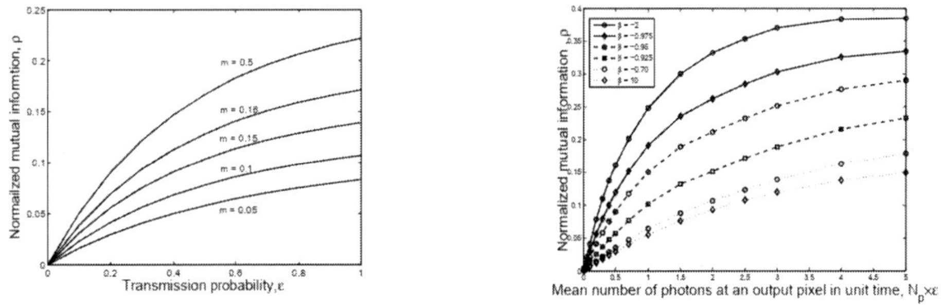

(a) (b)

Figure 1. Results for (a) the Markov chain model and (b) the Markov random field model.

We have also generated MRF images according to the generalized Ising model. We followed the Metropolis sampling algorithm [4] to generate 3-bit images of size 128×128 with varying spatial-correlation parameter, β. (The temperature parameter, T, is set to 3.) The algorithm is run for 1000 iterations for each image generated. We estimated the entropy and mutual information as in [3]. As in the case of the 1D Markov-chain model, we plot ρ as a function of transmission probability, ϵ, parameterized by β, as shown in Fig. 1(b). It is seen that ρ increases monotonically with ϵ. Moreover, for a fixed ϵ, ρ increases as β is decreased (i.e., as correlation metric is increased). This confirms the above conclusion that the effectiveness of the imaging PCII approach is enhanced with the presence of spatial correlation in the source image.

References

[1] S. Yeom, B. Javidi, and E. Watson, "Three-dimensional distortion-tolerant object recognition using photon-counting integral imaging," Optics Express **15**, 1513–1533 (2007).
[2] T. M. Cover and J. A. Thomas, *Elements of Information Theory* (John Wiley & sons, 1991).
[3] E. Volden, G. Giraudon, and M. Berthod, "Information in Markov random fields and image redundancy," in "Selected Papers from the 4th Canadian Workshop on Information Theory and Applications II," (Springer-Verlag, London, UK, 1996), pp. 250–268.
[4] S. Geman and D. Geman, "Stochastic relaxation, gibbs distributions, and the bayesian restoration of images," IEEE Trans. Pattern Analysis and Machine Intelligence **6**, 721–741 (1984).
[5] S. R. Narravula, M. M. Hayat and B. Javidi, "Information theoretic view of photon-counting integral imaging," submitted to Optics Express, (2009).

Combining Generalized Phase Contrast with matched filtering into a versatile beam shaping approach

J Glückstad[1] and D Palima
DTU Fotonik, Dept. Photonics Engineering
Technical Univ. Denmark, Ørsted Plads 343
DK-2800 Kgs Lyngby, Denmark

E-mail: jesper.gluckstad@fotonik.dtu.dk
URL: www.ppo.dk

Abstract. We adapt concepts from matched filtering to propose a method for generating reconfigurable multiple beams. Combined with the Generalized Phase Contrast (GPC) technique, the proposed method coined mGPC can yield dynamically reconfigurable optical beam arrays with high light efficiency for optical manipulation, high-speed sorting and other parallel spatial light applications [1].

1. Introduction

We consider a well-known and studied technique, optical matched filtering and propose an alternative way of utilizing it in combination with the Generalized Phase Contrast method [2]. Matched filtering and optical correlation is generally associated with pattern recognition applications that aim for detecting the presence of a target pattern within a given input scene [3]. In optical pattern recognition, a filter placed at the spatial Fourier plane generates a light spike where the image of a target pattern would otherwise be formed and centred when removing the filter. Conventionally, a correlation filter is designed and optimized to detect a specific pattern within an input scene. Since the user typically does not have any control over the spurious information contained within the scene, the optical correlation system is often faced with tradeoffs between various properties such as efficiency, output noise, input distortion and background noise robustness, and input discrimination ability, among others. Instead of passively waiting to detect the presence of a target pattern in an input scene, the user directly controls the input to generate a desired output. This involves choosing an optimal correlation target pattern and its matching filter and then arranging replicas at the correlator input to mimic the intended configuration of light spikes in the desired light projection pattern. A static element may be used for the filter and a dynamic input modulator may be used to reconfigure the output through straightforward and intuitive changes at the input. This is an advantage over using a single computer-generated diffractive element that always requires global input optimization even for simple changes at the output such as adding another spike, or repositioning a single spike. We show an experimental proof-of-principle of this so-called mGPC-based pattern projection and demonstrate its suitability for high-speed optical sorting.

---

[1] To whom any correspondence should be addressed.

2. The mGPC method

The mGPC method [4] combines the functions of a cascaded GPC with an optical matched filtering or optical correlator. The mGPC filter, $H(f_x,f_y)A^*(f_x,f_y)$, is determined by the product of the GPC filter, $H(f_x,f_y)$, and the correlation filter, $A^*(f_x,f_y)$. The process is equivalent to an expanded implementation, where a GPC output, $b(x,y)$, is optically correlated with another function, $a(x,y)$. Denoting the respective Fourier transforms of $a(x,y)$ and $b(x,y)$ as $A(f_x,f_y)$ and $B(f_x,f_y)$, it follows from Fourier transform properties that a cross-correlation $a(x,y)\star b(x,y)$ in the spatial domain corresponds to the product $A^*(f_x,f_y)B(f_x,f_y)$ in the Fourier domain. This is mathematically expressed as

$$\Im\{a(x,y)\star b(x,y)\} = A^*(f_x,f_y)B(f_x,f_y) \tag{1}$$

where $\Im$ represents the Fourier transform operation, $\star$ represents correlation and $A^*(f_x,f_y)$ is the complex conjugate of $A(f_x,f_y)$. This Fourier relation is the basis for the typical $4f$ optical correlation system known in matched filtering. The first Fourier lens projects the Fourier transform of the input plane field, $b(x,y)$ at the back focal plane where it is multiplied by $A^*(f_x,f_y)$ using the filter. The second Fourier lens generates the correlation $a(x,y)\star b(x,y)$ at the output plane (tractable coordinate inversions were neglected for simplicity).

The mathematical outline shows that an mGPC system exhibits linearity and space invariance. Space invariance implies that the output can be reconfigured through straightforward rearrangement of the patterns at the input. Thus, as in conventional GPC, dynamic reconfiguration in mGPC requires no computational overhead. This is an advantage in applications such as dynamic multiple beam steering, optical landscape projections and parallel optical sorting applications [5-6]. The mGPC can be further optimized according to relevant output metrics as shown in [4, 7]

References

[1] J. Glückstad, "Microfluidics: Sorting particles with light," Nature Materials **3,** 9–10 (2004).

[2] J. Glückstad and P.C. Mogensen, "Optimal phase contrast in common-path interferometry," Appl Opt 40, 268-282 (2001).

[3] A. Vander Lugt, "Signal Detection by Complex Spatial Filtering," IEEE Trans. Inf. Theory IT-10, 139–145 (1964).

[4] J. Glückstad, D. Z. Palima, J. S. Dam, I. Perch-Nielsen, "Dynamically reconfigurable multiple beam illumination based on optical correlation," J. Opt. A: Pure Appl. Opt. 11 034012 (2009).

[5] I. Perch-Nielsen, D. Z. Palima, J. S. Dam, J. Glückstad, "Parallel particle identification and separation for active optical sorting," J. Opt. A: Pure Appl. Opt. 11 034013 (2009).

[6] J. Glückstad, D.Z. Palima, J. S. Dam, I. Perch-Nielsen, "Parallel and real-time trapping, manipulating and characterizing microscopic specimens," (selected paper, Special issue: Optics in 2008), Opt. Photonics News 19 (12), 41 (2008).

[7] J. Glückstad, D.Z. Palima, Generalized Phase Contrast - Application in Optics and Photonics, Springer Series in Optical Sciences, Vol. 146, 310 pages (2009).

Journal of Physics: Conference Series 206 (2010) 012007 doi:10.1088/1742-6596/206/1/012007

Some topics in Quantum Imaging

L.A. Lugiato, A. Gatti, E. Brambilla, L.Caspani, P. Di Trapani, O. Jedrkiewicz, F. Ferri, D. Magatti

INFM-CNR-CNISM, Dipartimento di Fisica e Matematica, Universitá dell'Insubria, Via Valleggio 11, 22100 Como, Italy

E-mail: luigi.lugiato@uninsubria.it

Abstract. The field of Quantum Imaging exploits the quantum nature of light and the intrinsic parallelism of optical signals to devise novel techniques for optical imaging and for parallel information processing at the quantum level (see [1] and references quoted therein). In this presentation, we will shortly discuss two topics in this area: the first is the so-called ghost imaging which, however, is not necessarily quantum; the second is the detection of faint amplitude objects with a sensitivity beyond the standard quantum limit, and in this case we are fully in the quantum domain. Both topics are related to the phenomenon of optical parametric down-conversion (PDC), in which a fraction of the pump photons of a laser beam, injected into a crystal with a quadratic non-linearity, are down-converted to a pair of signal and idler photons, with conservation of total energy and total momentum. A feature of paramount importance is that the signal and idler beams are spatially correlated both in the near field (position correlation) and in the far field (momentum correlation).The simultaneous presence of position and momentum correlation implies quantum entanglement, as it has been also observed experimentally [2].

1. Ghost Imaging

This protocol was invented by Klyshko [3] as an application of the entanglement of signal and idler photons. The signal and idler beams identify a test and a reference arm (Fig.1). An object (e.g. a double slit in the figure) is located in the test arm. In order to obtain an image of the object, it would be necessary to use an array of detectors (e.g. a CCD camera), but one locates only a single detector in the test arm. On the other hand, one puts an array of detectors in the idler arm, but one does not get an image from this array either, because the idler photons do not interact with the object. The image of the object is obtained by detecting the coincidences between the arrival of a signal photon in the single detector and the arrival of an idler photons in one of the detectors of the array (Fig. 1). In this way, one exploits the position correlation between the signal and the idler photon. Important is also that, by simply modifying the configuration of the optical elements in the reference arm one obtains, instead of the image of the object, the Fourier transform of the intensity distribution of the object (the interference-diffraction pattern in the case of the double slit). In this case one exploits the momentum correlation between the signal and the idler photon.
The experimental demonstration of these concepts was reported in [4, 5]. With the new century, a very lively debate developed on the question whether quantum entanglement is necessary or not to realize ghost imaging (see [1] for references). Bennink et al. [6] first demonstrated experimentally that it is not necessary. The ideal configuration to realize ghost imaging

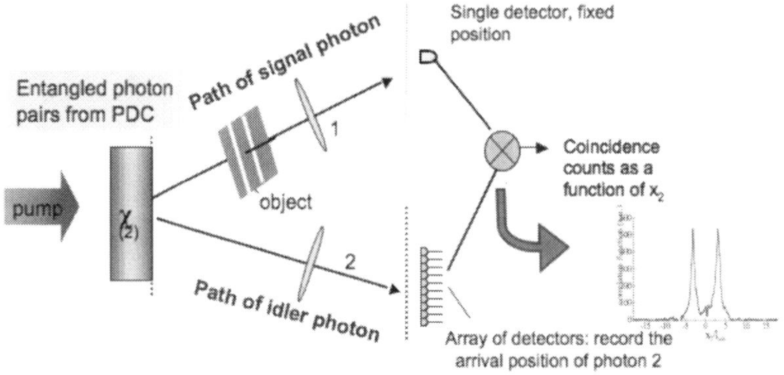

Figure 1. Scheme of ghost imaging.

(including the possibility of obtaining the Fourier transform of the intensity distribution of the object) by exploiting classical correlations instead of quantum entanglement is that of the so-called "thermal ghost imaging" [7, 8], see Fig. 2. A laser beam is injected into a rotating ground glass obtaining a pseudo-thermal beam, which is injected onto a 50/50 beam splitter. In this way one obtains two beams which are very well spatially correlated (even if not below the shot-noise level as it happens with the twin signal and idler beams emitted in PDC). Especially, these two beams are spatially correlated both in the near and in the far field, a condition which mimicks the simultaneous correlation in position and momentum of the signal and idler beams. One utilizes these two classically correlated beams exactly as the signal and idler beams in Fig. 1 and, as experimentally demonstrated in [8], one can well achieve all results of ghost imaging.

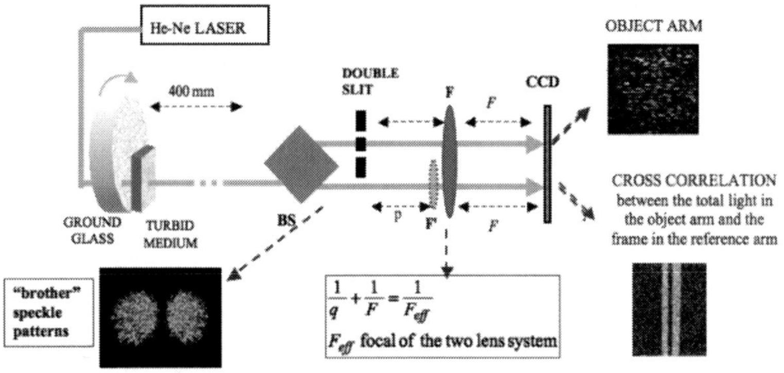

Figure 2. Thermal ghost imaging.

Recently Shapiro [9] has devised a variant of thermal ghost imaging, called computational ghost imaging, which allows to get rid of the reference arm.
In the recent trend towards applications of the ghost imaging protocol, there are therefore two options. One is to use quantum entangled beams generated by PDC; in this way one has the possibility of accessing the quantum regime. The other is to use pseudo-thermal beams, which are inexpensive and easy to handle.

2. Detection of faint amplitude objects

Let us consider first the problem of the detection of a weak absorption without considering, for the moment being, the spatial distribution of the absorber. If one wants to measure absorption with a high sensitivity, it is convenient to use the differential scheme shown in Fig. 3. The beam generated by any source is injected onto e 50/50 beam splitter, thus obtaining two classically correlated beams. The absorber is located in the path of one of the two beams, and the absorption is retrieved by measuring the difference between the photon numbers in the two beams after the absorption. The signal to noise ratio (SNR) in this measurement corresponds to the standard quantum limit (SQL) in such a measurement. Of course, the SNR can be increased by increasing the photon number in the two beams; however, if the absorber is delicate (e.g. it is a biological sample) one cannot increase the intensity arbitrarily and in this case it is convenient to be able to go beyond the SQL.

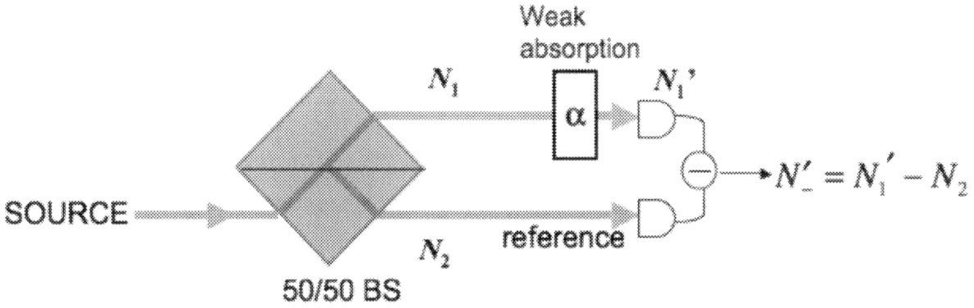

Figure 3. Differential scheme for the detection of a week absorption.

The SQL arises basically from the shot-noise level (SNL) in the fluctuations of the photon number difference. Therefore PDC offers the possibility of beating the SQL, since the fluctuations in the photon number difference can be below the SNL. Precisely, in Fig. 1 one replaces the source and the beam splitter by the pump laser and the nonlinear crystal of PDC and one uses the signal and idler beams instead of the two classically correlated beams. The experimental demonstration of the possibility of beating the SNR in this way was provided in [10, 11].Let us now consider the spatial distribution of the absorption. A natural idea is to generalize this approach to obtain the image of a weak amplitude object. In this case it is not possible to use spatially singlemode beams as those generated by the optical parametric oscillator utilized in [10, 11].
It is necessary to use spatially multimode signal and idler beams as those obtained in single-pass cavityless PDC. The experiment reported in [12] using cavityless PDC demonstrated that indeed in this way the fluctuations in the photon number difference between signal and idler can be below the shot noise level pixel by pixel, as it is necessary for the detection of a weak amplitude object with a sensitivity below the SQL.
However, the experimental setting of [12] is not appropriate for the detection of a weak amplitude object because the sub-shot-noise character arises only when the background noise of the CCD camera is subtracted. This problem has been thoroughly analyzed in [13], which demonstrates theoretically that the solution is to increase the duration of the pump pulses from the picosecond to the nanosecond range. In this regime the subtraction of the background noise becomes unnecessary, as it has been recently confirmed experimentally [14]. Now the experimental setting of [14] lends itself to the detection of faint amplitude objects.

References

[1] A. Gatti *et al.*, Quantum Imaging, in Progress in Optics Vol. **51**, p.251 (2008).
[2] J. Howell *et al.*, Phys. Rev. Lett. **92**, 210403 (2004).
[3] D.Klishko, Photons and Nonlinear Optics, Gordon and Breach (1988).
[4] T. Pittman *et al.*, Phys. Rev. A **53**, 2804 (1996).
[5] D. Strekalov *et al.*, Phys. Rev. Lett. **74**, 3600 (1995).
[6] R. Bennink *et al.*, Phys. Rev. Lett. **89**, 113601 (2002).
[7] A.Gatti *et al.*, Phys. Rev. Lett. **93**, 093602 (2004).

[8] F. Ferri *et al.*, Phys. Rev. Lett. **94**, 183602 (2005).
[9] J. Shapiro, Phys. Rev. A **78**, 061802 (2008).
[10] P. H. Souto Ribeiro *et al.*, Opt. Lett. **22**, 1893 (1997).
[11] Jiangrui Gao *et al.*, Opt. Lett. **23**, 870 (1998).
[12] O. Jedrkiewicz *et al.*, Phys. Rev. Lett. **93**, 243601 (2004).
[13] E. Brambilla *et al.*, Phys. Rev. A **69**, 053807 (2008).
[14] G. Brida *et al.*, Phys. Rev. Lett. **102**, 213602 (2009).

Journal of Physics: Conference Series **206** (2010) 012008 doi:10.1088/1742-6596/206/1/012008

Super-Resolved Imaging based upon Spatial Depolarization of Light

David Sylman[1], Zeev Zalevsky[1], Vicente Micó[2] and Javier García[2]

[1]School of Engineering, Bar-Ilan University, Ramat-Gan 52900, Israel

[2]Departamento de Óptica, Universitat de Valencia, c/Dr. Moliner, 50, 46100 Burjassot, Spain

E-mail: zalevsz@eng.biu.ac.il

Abstract. In this paper we present a new approach allowing the surpassing of the diffraction based limitation for the achievable resolution provided by imaging systems. It is based on an encoding-decoding process of various spatial pixels or regions in the field of view of the imaged object by orthogonal and differently time varying polarization states. The reconstruction of the original spatial information is obtained by applying a decoding process in a way similar to the encoding one. Although all the spatial information is summed and mixed together by the system, the decoding provides super resolved imaging since in every spatial position the undesired spatial information having time varying polarization dependence, that is uncorrelated to the decoding sequence applied on that specific spatial position, is averaged to zero and, on the other hand, the information which corresponds to that specific spatial region is being reinforced. The proposed approach can be used not only for super resolved imaging but also for imaging module that maintains the same spatial resolution while providing enlarged field of view

1. Introduction

Super resolution [1] is the field in which coding spatial degrees of freedom by other domains as time [2], wavelength [3], code or field of view [4], grey levels [5], coherence [6] and polarization [7] allows overcoming resolution limits of imaging systems posed by the diffraction effect. The spatial information is encoded by one of the other domains and after being transmitted through the imaging system it is decoded to reconstruct the high resolution image.

In recent work the coherence property of light was used to increase resolution or to obtain the same resolution over larger field of view [6]. In this work we follow the line presented in Ref. [6] while the capability for increasing the field of view is obtained not by encoding each region in the field of view by different axial delays (while each is larger than the coherence distance of the illumination source), but rather by proper source depolarization.

In the proposed concept the illumination source is depolarized in such a way that various spatial regions of the field of view are illuminated by differently time varying polarization states. The depolarization is such that the polarization variations in the different regions of the illumination spot are orthogonal to each other and thus time averaging may allow separating the various parts of the illumination spot (that coded different regions of the field of view) even if they were summed together. As in the case of field of view coding by the axial coherence of the illumination, here as well the time averaging duration that is required in order to orthogonally separate the multiplexed lateral regions is very short.

2. Experimental validation

In Fig. 1 one may see the setup for encoding/decoding of three spatial regions. In this case encoding polarizer 1 was fixed and polarizers 2 and 3 were rotating. The decoding polarizer 4 was rotating as well. Each rotating polarizer was shifted between 6 angular positions (each at angular difference of 30 degrees) while images were captured. Thus, for the encoding there were 36 (=6×6) different polarization states. The obtained results, after averaging all captured images, while applying in the decoding polarizer (polarizer 4) the required decoding sequence, are seen in Fig. 2.

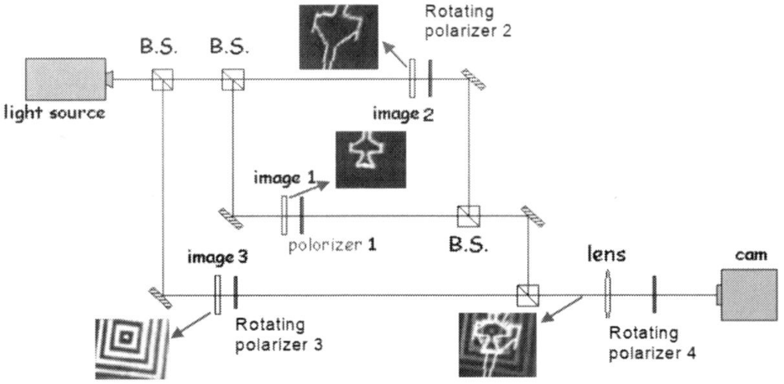

Figure 1. The experimental setup with three regions coding.

In Fig. 2(a)-(c) one may see that all three multiplexed images (coming from three different spatial regions 1, 2 and 3) were separated from each other. In Fig. 2(d) we repeat the same experiment when the image of region 2 is amplified 4 times more than the images of the rest of the regions while we try to reconstruct image from region 1 (the result of the reconstruction for the input of Fig. 2(d) is seen in Fig. 2(e)). One may see that although image from region 1 was attenuated more than the image from region 2, we yet could reconstruct it. This experiment shows that the orthogonal depolarization sequence allows reconstruction even when the required spatial information is lower in its SNR in comparison to the undesired information.

Figure 2. Reconstruction of image from: (a). branch 1. (b). branch 2. (c). branch 3. (d). Input branch 2 image is 4 times stronger than the rest of the images. (e). The reconstruction of image from branch 1 for input of Fig. 2(d).

References:

[1] Z. Zalevsky and D. Mendlovic "Optical Super Resolution," Springer (2002).
[2] M. Francon, *Nuovo Cimento Suppl.* **9**, 283-290 (1952).
[3] A. I. Kartashev, "Optical systems with enhanced resolving power," *Opt. Spectrosc.* **9**, 204-206 (1960).
[4] J. Solomon, Z. Zalevsky, and D. Mendlovic, "Geometrical superresolution using code division multiplexing," *Appl. Opt.* **42**, 32-40 (2005).
[5] Z. Zalevsky, P. García-Martínez and J. García, "Superresolution using gray level coding," *Opt. Express* **14**, 5178-5182 (2006).
[6] V. Mico, J. García, C. Ferreira, D. Sylman and Zeev Zalevsky, "Spatial information transmission using axial temporal coherence coding," *Opt. Lett.* **32**, 736-738 (2007).
[7] A. W. Lohmann and D. Paris, "Superresolution for nonbirefringent objects," *J. Opt. Soc. Am.* **3**, 1037-1043 (1964).

Hybrid imaging systems for depth of focus extension with or without post-processing

F. Diaz[1,2], F. Goudail[2], B. Loiseaux[1], J.-P. Huignard[1]

1 : Campus Polytechnique, 1 Avenue Augustin Fresnel, 91767 Palaiseau Cedex, France
2 : Laboratoire Charles Fabry de l'Institut d'Optique, CNRS, Universite Paris-Sud, Campus Polytechnique, RD 128, 91127 Palaiseau Cedex, France

E-mail: frederic.diaz@thalesgroup.com

Abstract. We address two different methods to enhance the depth of focus (DOF) according to the application, using amplitude and/or phase masks. The first application consists in increasing the length of the focal line of a lens while preserving its transverse resolution. We propose a mask inspired from holographic principles. We then address DOF enhancement with an hybrid imaging system composed of a pupil mask followed by a digital deconvolution step. We use an optimization criterion based on the final image quality to determine the optimal parameters of phase masks.

1. Introduction

Many types of pupil masks have been proposed for depth of focus (DOF) enhancement, with different properties and advantages which make them appropriate for different applications. The key point for choosing the optimal mask is to define the appropriate quality criterion related to the targeted application. The appropriate type of mask and the values of its parameters are then chosen to optimize this criterion. We will address two different types of applications. The first one consists in increasing the length of the focal line of a lens by using a novel type of mask inspired from holographic principles [1]. The second type of application concerns DOF enhancement in a hybrid imaging system composed of a pupil mask and a digital deconvolution step.

2. Uniform extended focal line: the holographically generated complex mask

The simplest designs to enhance the DOF are probably amplitude masks. Continuous phase masks and binary phase masks (BPM) can also be used, and they are usually considered more energy-efficient than amplitude masks since they have unit intensity transmission. However, in DOF extension, the relevant parameter is not the total transmitted energy, but the "useful" one, that is, the energy concentrated in the focal line. When this criterion is considered, amplitude/phase (complex) masks may have better performance.

We consider in this section that one wants to produce a light distribution as uniform as possible along the focal line. We define the depth of focus as the length of the region along the optical axis where the amplitude distribution has fluctuations lower than a given threshold. For given DOF and allowed fluctuations, our purpose is to determine the mask that maximizes the mean intensity in the DOF region. As an original alternative to the BPM [1], we propose a complex mask with the classical approach used for computer-generated holograms (Fig. 1). Such a design first specifies the desired normalized amplitude at chosen points along the optical axis, and then computes the mask that generates this distribution. This makes it possible to "tailor" the shape of the focal line. We will call this design "holographically generated complex mask" (HGCM) [2]. We achieved to obtain a DOF of $1.6mm$ with intensity fluctuation below 1% with a lens with a focal length $f = 160mm$, an aperture $D = 16mm$, at a wavelength $\lambda = 1064nm$ (Fig. 2).

The lower transmission of the HGCM (Fig. 1) is compensated by a better concentration of energy in the DOF region, thanks to the contribution of amplitude part of the filter. Moreover, with the HGCM, it is possible to achieve an accurate localization of the DOF region. In our example, the DOF region is centred on the focal point, whereas the DOF region of the BPM is asymmetrically distributed. Further advantages of the HGCM are the sharper decrease of the intensity at the limit of the DOF region, lower sidelobes and the better invariance of the transverse spot size with respect to z inside the DOF region.

The HGCM is thus a valuable alternative to the BPM for applications where very low fluctuations of the amplitude distribution along the optical axis are required and when the shape of the focal line has to be precisely tailored.

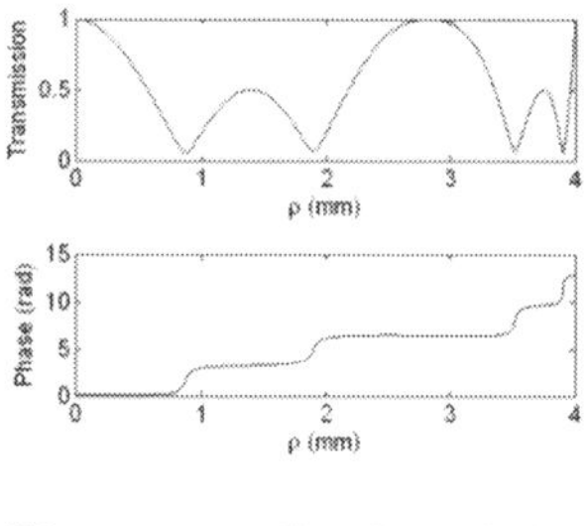

Figure 1. Profile of the HGCM

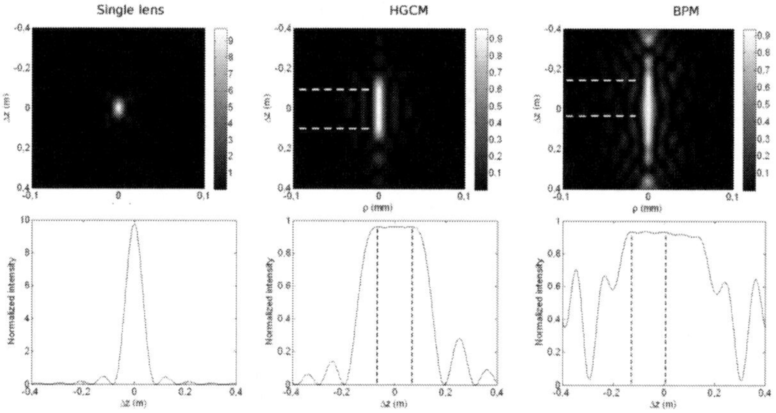

Figure 2. PSF of the lens, the BPM and the HGCM for a DOF of $1.6mm$ and fluctuations below 1%. The DOF region with variations below 1% are delimited with the dotted lines.

3. DOF extension with hybrid imaging system: simultaneous optimization of the deconvolution and the pupil mask

The approach discussed previously allows for moderate extension of the DOF while maintaining the transverse width of the point spread function (PSF) so that the resolution of the obtained image is little perturbed. When much higher DOF extension is required, the adequate masks tend to dramatically reduce the resolution. In this case, a deconvolution step is needed. In such hybrid imaging systems, two effects can degrade the quality of the final image: the variation of the PSF with respect to defocus, whereas the deconvolution filter is unique, and the noise enhancement due to deconvolution, which depends on the frequency response of the mask. Indeed, with usual phase masks, the more invariant to defocus the PSF is, the more they behave as low-pass filters. In such imaging systems, the relevant quality criterion is the quality of the final image obtained after the deconvolution step. We derived a mathematical expression of the output signal to noise ratio (SNR) that takes into account the detection noise and the defocus range where the system should provide an optimal image quality [3].

As an example, we consider the cubic phase mask [4] which is a pure phase mask whose PSF at the focal plane is $h(x,y) = |\text{FT}\{P(x,y)\exp[i\phi(x,y)]\}|^2$ with $\phi(x,y) = \alpha \times (x^3 + y^3)$, x and y the coordinates of the pupil, normalized to 1, $P(x,y)$ the pupil function, and α the phase excursion to be optimized. FT refers to the Fourier transform. The input SNR, defined by the SNR between the object and the noise of the sensor, is set at $SNR_{in} = 28dB$ and the maximal defocus value $\psi_{defocmax}$ is 9.2, with $\psi = \frac{\pi R^2}{\lambda}\left(\frac{1}{f} - \frac{1}{d_O} - \frac{1}{d_I}\right)$ where R is the aperture radius, f, d_O and d_I are respectively the focal length, the object distance and the image sensor plane distance. The object is the classical "Lena" image.

The value of SNR_{out} as a function of α is displayed in Fig 3. The optimal phase mask parameter is $\alpha_{opt} = 10$, leading to $SNR_{out} = 16.4dB$. With lower α, the PSF of the optical system is not insensitive enough to the defocus. With larger α, the PSF is much insensitive to defocus, but the optical system has then a lower-pass behaviour and the noise is enhanced by the deconvolution step. The parameter $\alpha_{opt} = 10$ is thus a compromise between theses two effects.

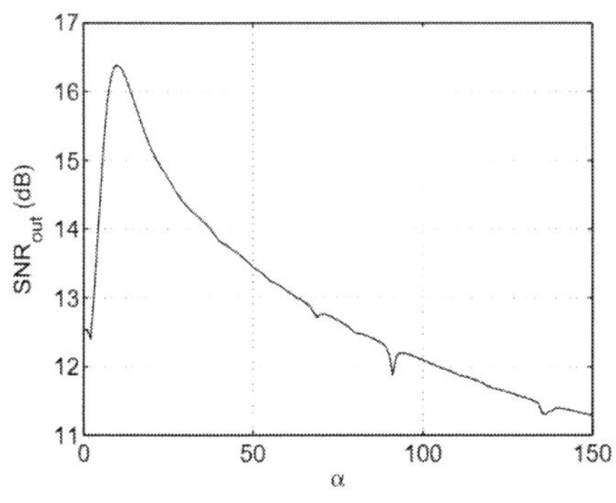

Figure 3. SNR_{out} in function of α for Lena.

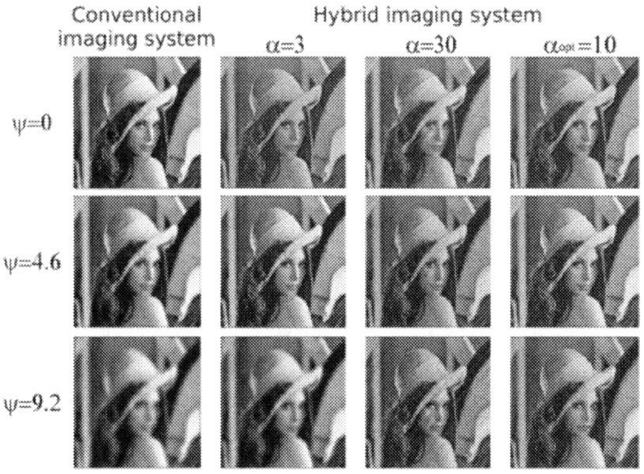

Figure 4. Results obtained with Lena with a conventional imaging system and with a hybrid imaging system at several values of α.

As shown on the deconvoluted images in Fig 4, the choosen criterion, SNR_{out}, has an effective correlation with the visual perception.

4. Conclusion

As the relevant optimization criterion strongly depends on the targeted application, we have discussed two approaches to DOF extension that are adapted to different cases. For applications where an intensity extended focal line is desired, an appropriate optimization criterion is the energy along this focal line. For imaging systems, the optimisation criterion should be the final image quality for which the SNR is an appropriate criterion.

References

[1] L. Liu, F. Diaz, L. Wang, B. Loiseaux, J.-P. Huignard, C. J. R. Sheppard, N. Chen, "Superresolution along extended depth of focus with binary-phase filters for the Gaussian beam", J. Opt. Soc. Am. A **25**, 2095 (2008)

[2] F. Diaz, F. Goudail, B. Loiseaux, J.-P. Huignard, "Design of a complex filter for depth of focus extension", Opt. Lett., **34**, 1171 (2009).

[3] F. Diaz, F. Goudail, B. Loiseaux, J.-P. Huignard, "Increase in depth of field taking into account deconvolution by optimization of pupil mask, **34**, (2009).

[4] W. Cathey, E. Dowski, "New paradigm for imaging systems", Appl. Opt, **41**, 6080 (2002).

3D-laser refractography – new chapter of information optics

B S Rinkevichyus[1], O A Evtikhieva, M V Yesin, I L Raskovskaya, A V Tolkachev
V A Fabrikant Physics Department, Moscow Power Engineering Institute,
Krasnokazarmennaya 14, Moscow, 111250, Russia

Abstract. The basic principles of laser refractography (LR), a new method for diagnostics of optically inhomogeneous media and flows, are described. This technology is based on the refraction of the structured laser radiation (SLR) in transparent inhomogeneity, digital registration and computer processing of the refraction pattern. The experimental setup allowed quantitative visualization of the inhomogeneous transparent media by using 2D or 3D-refractograms. The 3D-refractograms are used for diagnostics of the boundary layer near the hot or cold bodies in water are present.

1. Introduction

Laser refractography is a new technology of laser diagnostic for optically inhomogeneous media such as liquid, gas and plasma flows [1]. LR is based on refraction of structured laser radiation in optically inhomogeneous media. Structured laser radiation (SLR) is a spatially amplitude-modulated radiation obtained mainly with the help of Diffraction Optical Elements (DOE). The main types of SLR sources are classified according to the shape of 3D geometrical shapes formed by the beams from the sources of line-, plane- and cone-structured laser radiation. A 2D-refractograms [2] is the 2D-image of a source of spatially structured radiation formed by an optical system whose role is played by the medium under study. 3D-refractograms is a 3D-image of the surface formed by the rays from SLR refracted in the medium, and can be obtained from the entire body of experimental and theoretical 2D-refraction patterns for various cross sections with the help of a special technique.

2. Calculated SLR refractograms for a spherical thermal boundary layer

Note that all optical methods in fact describe the field of the refractive index, which can be subsequently transformed into the temperature field $T(x, y)$ or the field of some other parameter [3]. Following, we shall analyze thermal processes assuming that for laser wavelength 0.6328 μm, the temperature dependence of the refractive index of water is determined by the approximate dependence $n(t)$. The mathematical simulation of refraction of a LSR in a spherically symmetric thermal boundary layer around a sphere of radius R is performed for the radial temperature dependence described by the expression

$$T(r) = T_0 + \Delta T \exp\left(-(r-R)^2/a^2\right), \tag{1}$$

where $T_0, \Delta T$ and a are model parameters. Figure 1 shows calculated 3D-refractogram of the cylindrical SLR for temperature distribution (1) in the vicinity of a hot ball in water at $T = 90$ °C, $R = 20$ mm.

---

[1] Please address correspondence to Prof. Rinkevichius B S, rinkevbs@mail.ru.

3. Experimental 3D-refractograms receiving

The method of 3D image reconstruction was originally developed in [4]. Hear it applied to reconstruct 3D refractograms by number of 2D cross-sections.

The refractography experiment illustrates the process of hot metal ball natural cooling in cold water. Semiconductor laser radiation with wave length 0.670 μm passing through DOE element witch makes the laser sheet and then under the hot metal ball. Laser sheet thickness under the ball was 0.1 mm, width was 51 mm, distance from the centre of laser sheet and ball was 0.075 – 0.1 mm. By the metal ball cooling the shape of laser sheet become straight. It is possible to reconstruct 3D spatial-temporary image by the number of frames registered on the stationary semitransparent screen. This image will illustrate water refraction index alteration caused by transitional temperature field gradient near the hot ball. Figure 2 shows the registered frame sequence. Time-spatial 3D refractogram reconstructed by 80 frames is shown in figure 3.

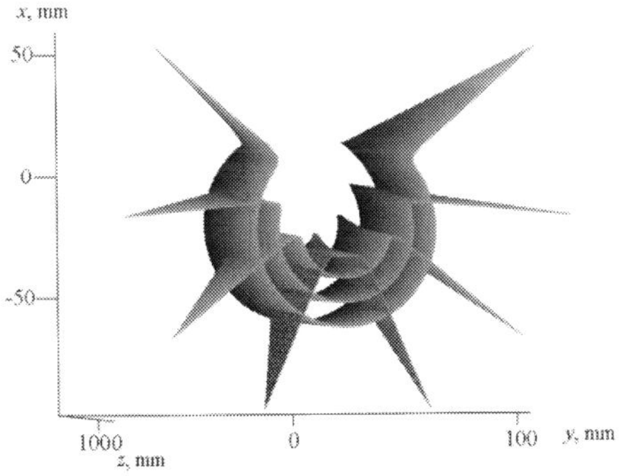

Figure1. Calculated 3D-refractograms for a cylindrical SLB propagating in water under a hot ball.

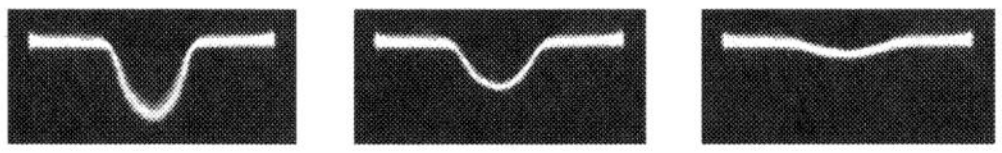

Figure 2. The registered frames.

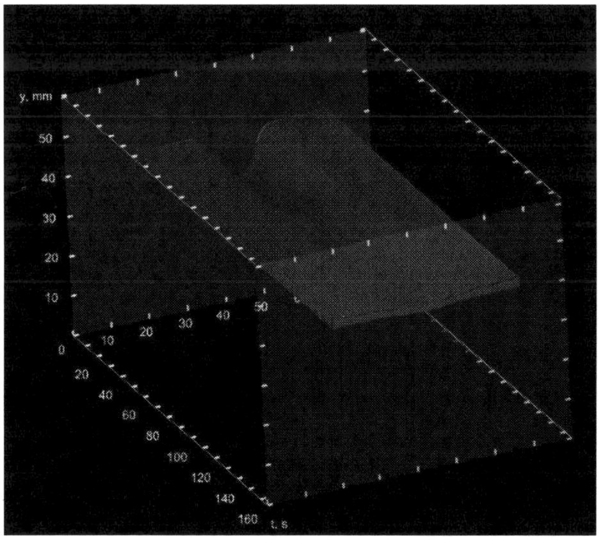

Figure 3. Time-spatial 3D refractogram reconstructed.

4. Conclusion

The new technology laser refractography described in the report uses SLR for obtaining 2D- and 3D-refractograms of transparent inhomogeneous media or flows. In a certain sense, a refractogram is a "visualized portrait" of the medium under study and can be compared with the set of elementary refractograms of typical inhomogeneities. This makes it possible to perform a quick diagnostics of the medium right in the course of observations.

5. References

[1] Evtikhieva O A, Raskovskaya I L and Rinkevichyus B S 2008 *Laser refractography* Moscow, Phizmatlit, 176 p. (in Russian).

[2] Raskovskaya I L, Rinkevichyus B S and Tolkachev A V 2007 Laser refractography of optically inhomogeneous media *Quantum electronics* **37** **12** 1176-1180

[3] Rinkevichius B S 1998 *Laser diagnostics in fluid mechanics* New-York, Begell House, Inc. 340 p.

[4] Yesin M V, Rinkevichius B S and Tolkachev A V 2002 3D Images Reconstruction of the Objects with Indistinct Boundaries *Proc. SPIE* **4900 Part Two** 1140-1146.

Optical-sectioning microscopy by patterned illumination

G Saavedra, M Martinez-Corral, E Sanchez-Ortiga and A Doblas
3D Display & Imaging Group, Universitat de València (Spain)
E-mail: genaro.saavedra@uv.es

Abstract.
We propose a very simple method for the flexible production of 1D structured illumination for high resolution 3D microscopy. Specifically, we propose the insertion of a Fresnel biprism after a monochromatic point source for producing a pair of twin, fully coherent, virtual point sources. The resulting interference fringes are projected into the 3D sample and, by simply varying the distance between the biprism and the point source, one can tune the period of the fringes, while keeping their contrast, in a very versatile and efficient way.

1. Variable-frequency structured illumination

Non-uniform illumination techniques are between the most promising proposals to overcome the inherent limitations of conventional light microscopes regarding both lateral spatial resolution and optical sectioning capability[1]. In structured illumination (SI) microscopy a transverse periodic 1D pattern is projected onto the specimen and a stack of 2D images is recorded after scanning the object axially [2, 3, 4]. This illumination structure acts as a *carrier* periodic pattern, which generates several spectrally-shifted replicas of the original spatial-frequency content of the object. Through a phase-shift method, an extended spectrum of the object can be recovered over the image. This completion not only extends the lateral spatial resolution but also provides the system with optical sectioning capabilities.

From a practical point of view, in the classic SI microscopy an illumination plane wave is splitted into two tilted ones by means of a 1D diffraction grating. Their interference generates the desired periodic pattern inside the 3D sample under issue. It is worth to note, however, that different goals in SI microscopy —optimum optical sectioning or maximum lateral resolution— cannot be achieved simultaneously, and require different values of the frequency of the illumination pattern. Moreover, this frequency depends not only on the period of the *splitting* grating, but also on the optical power of the microscope objective. Thus, a change in this inspection objective would require a readjustment of the period of the grating to achieve the proposed optimization of the features of the SI microscope. This desirable tunability is not easily reached in the classic SI scheme, unless sophisticated variable-frequency gratings are used.

Note that a pair of interfering plane waves can be also obtained from two fully coherent point sources *collimated* by an spherical lens. In this case, the angle between the exiting waves is controlled by the focal length of the lens and, remarkably, by the distance between the point sources. Our proposal is to generate these fully coherent twin sources by use of a Fresnel biprism. When this element is illuminated by a point source, two virtual point sources are generated on the same transverse plane as the original one, symmetrically located around it, with a separation proportional to the distance between the biprism and the real source. Thus, by simply shifting the biprism respect to the real point source one can change the separation between the virtual sources. If we consider now an spherical lens with its back focal plane laying on the virtual twin sources location, a pair of interfering plane waves is generated, whose relative angle can be varied simply by changing the gap between the sources. This tunability can be added to the classic SI microscope by replacing the grating and the plane-wave illumination by a point source, a Fresnel biprism and a collimating lens. The axial displacement of the biprism in this new setup generates a continuous variation in the frequency of the 1D periodic illumination pattern into the sample. Note that our proposal implies very small changes in the classic scheme of the SI microscope, in such a way that its implementation is suitable even into commercially available SI instruments.

As a test for our proposal, we use implement a tunable-frequency SI microscope by use of a Fresnel biprism with refringence angle $\delta = 0.5°$ and refraction index $n = 1.51$. We generate the real point source from the output of an optical fiber coupled with a standard HeNe laser ($\lambda = 632.8$ nm). For the collimation of the twin spherical wavefronts we use an achromat with focal length $f = 200$ mm. This setup was followed by a standard microscope with a $f = 150$ mm tubelens and a low-NA (0.10) low-magnification ($5\times$) objective.

We present in Fig. 1(a) and Fig. 1(b) two images of the illuminated sample plane for two different axial positions of the biprism. As a target we use a *USAF 1951* test, that allows a direct comparison of the periodic illumination and the callibrated spatial-frequency sections of the test. A variation by a factor of 5 in the illumination spatial-frequency is clearly shown in this two pictures. We use also the highest of these *carrier* spatial-frequencies to obtain a superresolved synthetic image of the target under issue. Figures 1(c) and (d) show, respectively, the original and synthetic images obtained in this case, showing clearly the improvement in the lateral resolution predicted by the SI technique.

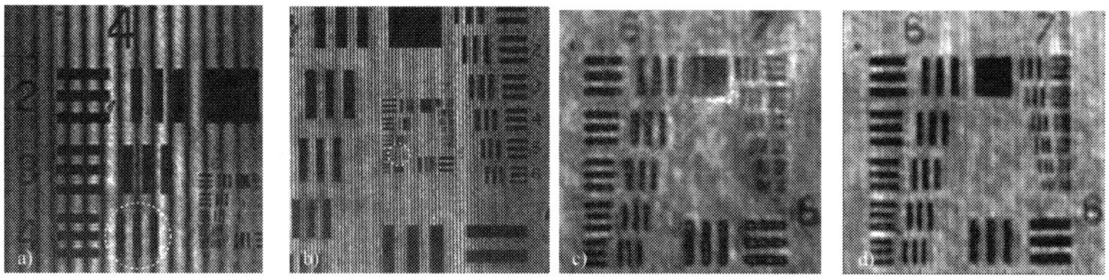

Figure 1. Experimental results for the the tunable-frequency SI microscope: (i) Illuminated target for two different axial locations of the biprism; the dashed circles correspond to the groups in the target with: (a) 22 lp/mm; (b) 102 lp/mm. (ii) Images of the central region of the target: (c) Original brightfield setup; (d) SI synthetic image with the illumination pattern in (b).

Acknowledgments

This work has been funded by the Plan Nacional I+D+I (grant DPI2006-8309), Ministerio de Educación y Ciencia (Spain).

References

[1] Martínez-Corral M and Saavedra G 2009 *Prog. Opt.* **53** 1–68
[2] Lukosz W 1966 *J. Opt. Soc. Am.* **56** 1463–1471
[3] Neil M, Juskaitis R and Wilson T 1997 *Opt. Lett.* **22** 1905–1907
[4] Gustafsson M 2000 *J. Microsc.* **198** 82–87

Biological applications of novel nonlinear optical microscopy

Shin'ichiro Kajiyama[1, 2], Yasuyuki Ozeki[2], Kiichi Fukui[3] and Kazuyoshi Itoh[2]

[1] School of Biology-oriented Science and Technology, Kinki University, 930 Nishimitani, Kinokawa, Wakayama, 649-6493, Japan

[2] Department of Material and Life Science, Graduate School of Engineering, Osaka University, 2-1 Yamadaoka, Suita, Osaka 565-0871, Japan

[3] Department of Biotechnology, Graduate School of Engineering, Osaka University, 2-1 Yamadaoka, Suita, Osaka 565-0871, Japan

E-mail: kajiyama@waka.kindai.ac.jp

Abstract. Two types of newly developed nonlinear optical microscopes namely stimulated parametric emission (SPE) microscope and stimulated Raman scattering (SRS) microscope were presented together with their biological applications.

1. Introduction

The wave of nonlinear coherent microscopy (NLCM), such as coherent anti-Stokes Raman scattering (CARS), second-harmonic generation (SHG) and third-harmonic generation (THG) in recent years is a response in quest of label-free/non-staining 3D-microscopy. Using sub-picosecond pulsed laser, these techniques enabled high special resolution and non-invasive imaging of labile samples such as living cell Since each NLCM has its own contrast mechanism and provides different optical images, it is important and meaningful to find a novel process in NLCM that gives a new optical image based on a new contrast source. In this workshop the authors present two kind of newly developed nonlinear optical microscopes, namely stimulated parametric emission (SPE) microscope and stimulated Raman scattering (SRS) microscope together with their biological applications.

2. Stimulated parametric emission (SPE) microscopy

Stimulated parametric emission (SPE) microscopy is a new technique that has been realized by our group 30 years after the theoretical prediction[1] of its possibility. SPE is a four-wave mixing (FWM) process in which the mixing of two frequencies at ω_1 (pump) and at ω_2 (probe) in sample produce a signal at $\omega_{SPE} = \omega_1 + \omega_1 - \omega_2$ when the phase matching condition is satisfied. The SPE microscopy shows its possibility to detect various kinds of materials in principle[2,3] due to refractive index change and electronic resonances in the samples. In our experiment, a mode-locked Ti : sapphire laser with wavelengths ranging from 670 to 950 nm pumped by a continuous-wave laser at 532 nm was employed. A comparison between bright-field and SPE images of tobacco BY-2 cultured cell demonstrated that the SPE image showed detailed inner structure of the cells with a distinct nucleus, endoplasmic reticulum and a matrix of microtubules that were not visualized clearly by conventional optical microscopy without staining (Fig. 1a,b).

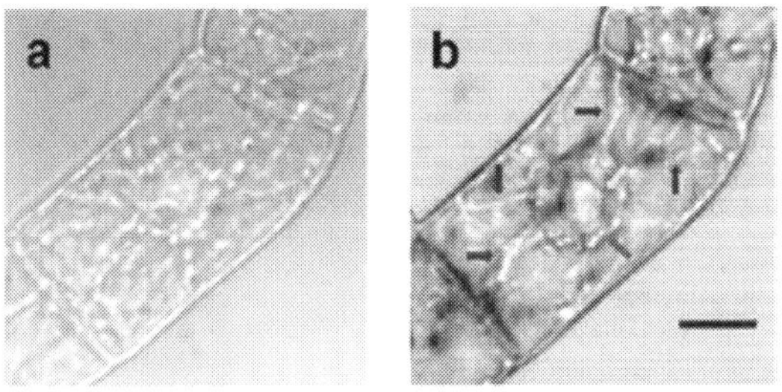

Figure 1. Comparison between Bright-field image (a) and SPE image (b) of an unstained BY-2 cell. Bar: 15 μm.

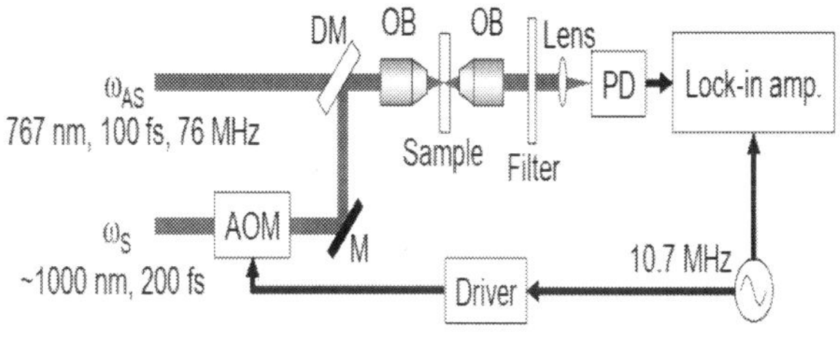

Figure 2. Experimental set-up of SRS microscopy. (DM: dichroic mirror, OB: objective lens, PD: silicon photodiode, AOM: acousto-optic modulator)

3. Stimulated Raman Scattering (SRS) microscopy

SRS is a well-known nonlinear Raman phenomenon in which photon energy is transferred from an anti-Stokes (AS) pulse to a Stokes (S) pulse when the frequency difference matches a Raman vibrational mode[4]. Similarly to CARS, SRS originates from 3rd-order nonlinear susceptibility $\chi^{(3)}$. SRS reflects only vibrational resonances (i.e. Im $\chi^{(3)}$) and is unsusceptible to nonresonant nonlinearities (i.e. Re$\chi^{(3)}$), providing us high vibrational contrast. Moreover, we pointed out in Ref. 5 that the theoretical sensitivity limit of SRS microscopy is comparable to that of CARS microscopy because SRS can be viewed as the homodyne detection of the nonlinear optical signal with the excitation pulse. In order to detect the small intensity change of AS pulse, the S pulse was intensity-modulated in time, and the SRS signal was obtained through the lock-in detection of the intensity of the AS pulse in our setup shown in Fig. 2.

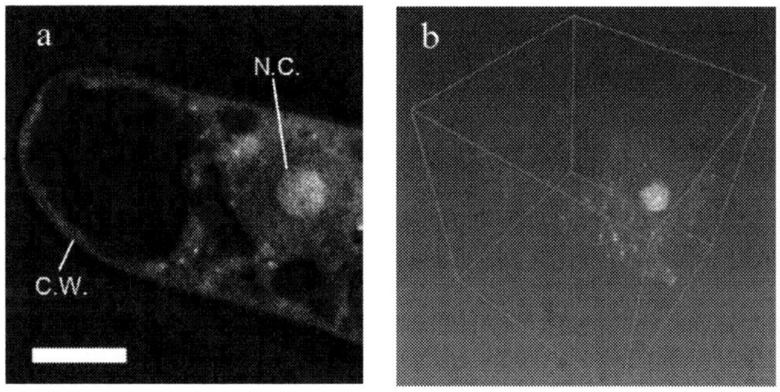

Figure 3. SRS images of an unstained tobacco BY-2 cell.
(C.W.: cell wall, N.C.: nucleus, Scale bar: 10 μm.Raman shift: 2967 cm^{-1})

SRS microscopy is readily applicable to the observation of unstained living cells. Considering that the lock-in signal is proportional to Im $\chi^{(3)}$, we constructed images in such a way that the image intensity is proportional to (Im $\chi^{(3)})^2$. As shown in Fig. 3(a)(b), nuclei and cell walls of a tobacco BY-2 cultured cell were clearly visualized in 2D and 3D, based on vibrational contrast by CH stretching modes.

4. Conclusion

In summary, SPE and SRS microscopy allow high-resolution imaging of unstained live-cells. The advantage of SPE microscopy over other nonlinear techniques is that it enables to gain contrast from all kinds of materials in principle. As for the SRS microscopy, we have demonstrated high-contrast vibrational imaging by suppressing the nonresonant background. This novel microscopy technique would be advantageous especially in high-contrast molecular mapping of weak substances in solvent, which is typical in the observation of biological samples. All these features proved that SPE and SRS microscopy would be new and effective tools for life science and biotechnology.

References

[1] Levenson M D, Bloembergen N 1974 *Phys. Rev. B* **10**, 4447-63
[2] Isobe, K. et al. 2006 *Opt. Express* **14**, 786-93
[3] Isobe, K. et al. 2006 *Opt. Express* **14**, 11204-14
[4] Boyd, R. W. 2003 '*Nonlinear optics*,' 2nd ed., Academic Press
[5] Ozeki Y. et al. 2009 *Opt. Express* **17**, 3651-58

Diversity of optical signal processing led by optical signal form conversion

Tsuyoshi KONISHI

Osaka University, 2-1 Yamadaoka, Suita Osaka 565-0871, JAPAN, .

E-mail: konishi@mls.eng.osaka-u.ac.jp

Abstract. This paper reviews opportunities of optical signal form conversion as typified by time-space conversion in optical signal processing. Several examples of typical ultra-fast optical signal processing using optical signal form conversion are described and their applications are introduced in respect to photonic networks, ultra-fast measurement, and so on.

1. Introduction

Optical technology is expected to play an important role in the current and next generation industrial or scientific challenges. To make a full use of the ability of optical technology in the new stage, it is necessary to promote a fundamental review of the nature of light with taking advantage of the features offered by current existing technology. Here, let me remember our original purpose and consider the meaning of time-space conversion again. Basically, a light wave can be represented by spatiotemporal parameters based on a wave-equation and we can use its spatiotemporal parameters for optical signal processing. From the theoretical viewpoint, a wave equation can be modified into its simplified versions in time and space domains.[1] They are respectively given by,

$$\left(\frac{\partial}{\partial z} - \frac{ik}{2}\cdot\frac{\partial^2}{\partial t^2}\right)\varphi = 0 \tag{1}$$

$$\left(\frac{\partial}{\partial z} + \frac{i}{2k_0}\cdot\Delta_\perp\right)\varphi = 0 \tag{2}$$

As you can see from these two equations, temporal and spatial behaviors of light have some similarities each other. It suggests that there is a possibility of exchange between temporal and spatial phenomena.

From the phenomenalistic viewpoint, wavelength is the shortest length which can be represented by a light wave and it is at most sub micrometers. In spatial domain, nanotechnology enables us to easily fabricate various dimensional structures in the range from several nanometers to several micrometers. Since this range includes a sub wavelength dimension, it enables us to design artificial materials such as a photonic crystal and control a light wave more leeway. This suggests that spatial dimensions of an ultrashort pulse and an optical device come to be overlapped and ultrafast photonics and nanotechnology could go well together there. Thus, physical advancements in optical technologies encourage various kinds of physical parameters to have close interactions between them.

In this paper, we review opportunities of optical signal processing assisted by optical signal form conversion for the next generation of photonics.[2-5]

2. Diversity of optical signal processing led by optical signal form conversion

From the viewpoint of optical signal processing, the manner of encoding based on intensity modulation is not always suitable for processing with optical technology. In this sense, it can be said that the true necessity is not optical replacement. One promising approach appears to be employing an appropriate interface for optical signal-form conversion among various physical parameters of optical signals to select an appropriate physical parameter for optical signal processing. Here, we show you representative example concerning optical A/D conversion.[6] The proposed system realizes optical quantization and optical coding after the optical sampling. In optical quantization, the power of the input analog signal is converted into the extent of the center-wavelength shift by use of self-frequency shifting in a fiber. Using the difference of the center wavelength, we can achieve optical quantization of an input analog signal. Thus, we can achieve photonic A/D conversion using amplitude-wavelength-space-time conversions.

On the other hand, current progress in ultrafast photonics and nanotechnology suggests that they could go well together. In addition, a wave equation can be simplified to these two equations in spatial and time domain as shown in Equations (1) and (2). This suggests the spatio-temporal analogy and exchangeability between them and the derivatives in equations are 2 dimensional ones. There is now much research proceeding in this topics. For example, we have proposed new approach for ultrafast optical time-resolved spectroscopy[7,8], 2-D pulse shaping[9,10], optical distortion equalizer[11].

3. Conclusion

This paper reviewed opportunities of optical signal form conversion as typified by time-space conversion in optical signal processing for the current and next generation of photonics. There is now much research proceeding in this field. For example, we have proposed new approach for optical label recognition[12,13], and digital-to-analog (D/A)[14] conversions, and optical limiter[15].

References

[1] S. A. Akhmanov, et al., Optics of Femtosecond Laser Pulses: (AIP press, New York, 1992), Chap. 1.
[2] T. Konishi and Y. Ichioka, J. Opt. Soc. Am. A Vol.16,1076 (1999).
[3] T. Konishi and Y. Ichioka, Optical Information Processing: A tribute to Adolf Lohman (SPIE press, 2002), Chap. 14.
[4] T. Konishi, OFC/NFOEC2005, OTuG1 (invited paper, 2005).
[5] T. KONISHI, Lasers, Optics and Electro-Optics Research Trends: (Nova Science Publishers, New York, 2007) Chap.1.
[6] T. Konishi, et al., J. Opt. Soc. Am. B, Vol. 19, 2817 (2002).
[7] T. Konishi and Y. Ichioka, Optical Review, Vol. 6, 507 (1999).
[8] K. Tanimura, et al.,, Opt. Rev., Vol. 10, 77 (2003).
[9] T. Konishi, et al., IEEE Photon. Technol. Lett., Vol. 16, 620 (2004).
[10] Y. Oshita, et al., IEEE Photon. Technol. Lett., Vol. 16, 623 (2004).
[11] T. Konishi et al., IEEE J. of Lightwave Technol. Vol. 24, 2693 (2006).
[12] T. Konishi, et.al., Opt.Lett., Vol. 18, 1445 (2001).
[13] H. Furukawa, et al., Electron. Lett., vol.40, pp.1440-1441, (2004).
[14] T. Nishitani, T. Konishi, and K. Itoh, Optics Express, Vol. 13, 10310 (2005).
[15] H. Goto, T. Konishi, and K. Itoh, J. Opt. A: Pure and Appl. Opt. 10, 095306 (4pp), (2008).

Generation of mid/far-infrared ultrashort pulses in organic crystals

K. Kuroda[1], Y. Toya[1], T. Satoh[1], T. Shimura[1], S. Ashihara[2], Y. Takahashi[3], M. Yoshimura[3], Y. Mori[3] and T. Sasaki[3]

[1]Institute of Industrial Science, University of Tokyo, Komaba 4-6-1, Meguro-ku, Tokyo 153-8505 Japan
[2]Department of Applied Physics, Tokyo University of Agriculture and Technology
[3]Department of Electrical Engineering, Graduate School of Engineering, Osaka University

E-mail: kuroda@iis.u-tokyo.ac.jp

Abstract. Far-infrared ultrashort-pulse in the wavelength range of 10 – 30 μm is generated using an organic nonlinear crystal DAST. The frequency down conversion is performed by the difference frequency generation between the signal and idler waves from an optical parametric amplifier pumped by the 120 fs Ti: sapphire laser pulses. The pulse width is measured using the cross-correlation with the pump. In the present paper, the shortest pulse width of 210 fs was obtained with the spectrum range of 10 – 30 THz.

1. Introduction

Ultra-short pulses in mid- and far-infrared region of spectrum are quite useful for investigating the dynamics of molecular vibrational structures, since a variety of vibrational modes are distibuted in this spectral range. They are also useful for the study of the dynamics of low-energy excitations in solid state materials.

Nonlinear optical frequency down conversion has been used for mid/far-IR short-pulse generation [1]. Some innorganic crystals, such as $AgGaSe_2$ and GaSe, are efficient for the generation of mid/far-IR pulses, but transparent range of these materials is limited down to 20 μm. Among many nonlinear optical crystals, an organic nonlinear optical crystal DAST (4-dimethylamino-N-methyl-4-stilbazolium tosylate) has been proved quite efficient for the frequency conversion down to THz wave region [2, 3, 4, 5]. In the present paper, we have demonstrated the generation of far-IR short-pulse in spectral range of 10 – 30 μm (10 – 30 THz) using DAST crystals and have measured the spectrum and the pulse width.

2. Experimental setup and results

Experimental setup for the generation of mid/far-IR short-pulses is shown in Fig.1. A β-barium borate-based optical parametric amplifier is pumped by a Ti:sapphire laser. The center frequency of the pump laser is located at 790 – 800 nm with the pulse width of 120 fs and the repetition frequency of 1 kHz. The wavelength of the signal beam from OPA is tuned between 1500 and 1590 nm, whereas that of the idler is from 1630 to 1710 nm. The pulse widths of the signal and idler are about 100 fs. The signal and idler are mixed in a 0.2-mm thick DAST crystal and the difference frequency wave is generated in the spectral range of 10 – 30 μm. The signal and idler beams are polarized along the a axis of the crystal so that the largest component of nonlinear optical coefficients d_{111} is available. The spectrum of the far-IR pulse is measured using the Fourier transform infrared spectrometer (FTIR) and the pulse duration is estimated by the cross-correlation with the pump pulse.

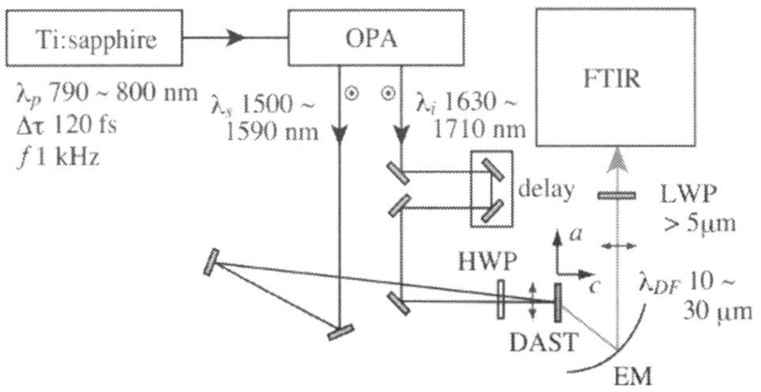

Figure 1. Setup of difference frequency generation. OPA: optical parametric amplifier, HWP: half-wave plate, EM: ellipsoidal mirror, LWP: long-wave pass filter, FTIR: Fourier transform infrared spectometer

Typical experimental results are shown in Fig.2. In order to satisfy the phase matching condition, non-collinear geometry is adapted with the non-collinear angle of 0.7°. Spectrum spreads over the region of 10 - 30 THz. There are several missing frequencys due to the absorption of DAST crystal. The pulse width is 210 fs and the energy is 310 nJ.

In summary, we have generated ultra-short mid- and far-infrared pulses by use of organic nonlinear crystal DAST.

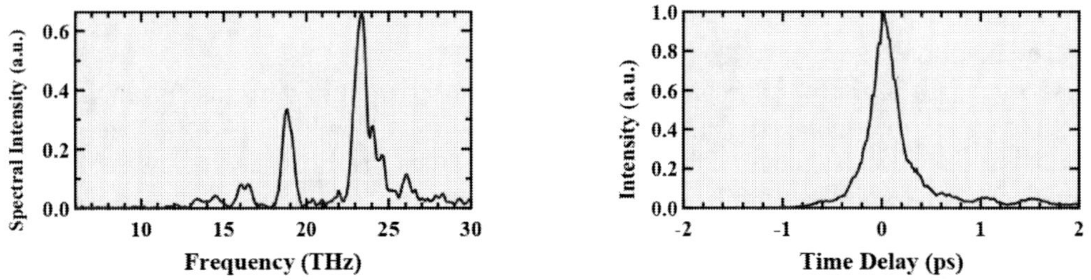

Figure 2. Experimental results. Spectrum (left) and ross-colliration trace (right).

References

[1] R. A. Kaindl, M. Wurm, K. Reimann, P. Hamm, A. M. Weiner, and M. Woerner 2000 *J. Opt. Soc. Am.* B **17**, 12

[2] K. Kawase, M. Mizuno, S. Sphma, H. Takahashi, T. Taniuchi, Y. Urata, S. Wada, H. Tashiro, and H. Ito 1999 *Opt. Lett.* **24**, 1065

[3] H. Adachi, T. Taniuchi, M. Yoshimura, S. Brahadeeswaran, T. Higo, M. Takagi, Y. Mori, T. Sasaki, and H. Nakanishi 2004 *Jpn. J. Appl. Phys.* **43**, L1121

[4] A. Schneider, M. Neis, M. Stillhart, B. Ruiz, R.U. A. Khan, and P. Günter 2006 *J. Opt. Soc. Am.* B **23**, 9

[5] Y. Takahashi, H. Adachi, T. Taniuchi, M. Takagi, Y. Hosokawa, S. Onzuka, S. Brahadeeswaran, M. Yoshimura, Y. Mori, H. Masuhara,T. Sasaki, H. Nakanishi 2006 *J. Photochem. Photobiology* A **183**, 247

Detection in polarimetric images in the presence of additive noise and non-uniform illumination

A. Bénière[1,2], F. Goudail[1], M. Alouini[3] and D. Dolfi[2]

[1] Laboratoire Charles Fabry de l'Institut d'Optique, CNRS, Univ Paris-Sud, Campus Polytechnique, RD 128, 91127 Palaiseau, France

[2] Thales Research and Technology - France, 1 Av. Augustin Fresnel, 91767 Palaiseau Cedex, France

[3] Institut de physique de Rennes, CNRS, Univ Rennes 1, 35042 Rennes, France

E-mail: arnaud.beniere@institutoptique.fr

Abstract. Active polarimetric imaging systems yield information about the intensity contrast and the Orthogonal State Contrast (OSC) in the scene. However, in real systems, the illumination is often spatially or temporally non uniform which creates artificial intensity contrasts that can lead to false alarms. We derive the Generalized Likelihood Ratio Test (GLRT) detectors when intensity information is taken into account or not. These results are used to determine in which cases considering intensity information in addition to polarimetric information is relevant or not.

A simple configuration consists in illuminating the scene with a totally polarized beam (whose state of polarization is located anywhere on the Poincaré sphere) and in acquiring two images: the first one X is formed with the fraction of the light having the same state of polarization as the illumination and Y is formed with the fraction of the light in the orthogonal state. We consider homogeneous sub-samples X_i and Y_i, $i \in [1, N]$ of the images, that may be spatial or temporal. The average number of photo-electrons under unitary and uniform illumination, denoted m_X and m_Y, are identical for all pixels. The intensities for the two polarization directions at each pixel $i \in [1, N]$ can be written as:

$$X_i = F_i\, m_X + n_i^x \quad \text{and} \quad Y_i = F_i\, m_Y + n_i^y, \tag{1}$$

where $\mathbf{F}$ denotes the spatial fluctuations of the illumination intensity. In other words, F_i represents the intensity of the incoming light and m_X (m_Y) represent the reflectivity of the observed material with respect to each polarization. One can use another parametrization where $I = m_X + m_Y$ is the intensity and $P = (m_X - m_Y)/(m_X + m_Y)$ the Orthogonal State Contrast (OSC). Each measure are perturbed by additive white Gaussian noises n_i^x and n_i^y with zero mean value and variance σ^2, which are assumed statistically independent.

To address detection, we will use the formalism explained in detail in Ref.[1]. The sample is divided into two parts $\Omega_a = [1, N_a]$ and $\Omega_b = [N_a + 1, N]$ so that $N_a + N_b = N$. The region a corresponds to an homogenous part of the image and so does region b. The first part of the subsample $\chi_a = \{X_i, Y_i, i \in \Omega_a\}$ is defined by parameters (I_a, P_a). The second part of the subsample $\chi_b = \{X_i, Y_i, i \in \Omega_b\}$ is defined by the parameters (I_b, P_b). We denote $\mathbf{F_a} = [F_i]_{i \in \Omega_a}$ and $\mathbf{F_b} = [F_i]_{i \in \Omega_b}$. Our objective is to know whether regions a and b have different polarimetric properties or not.

For that purpose, we have derived the optimal detection algorithm which is the generalized likelihood ratio test [2]. We first assume that the illumination pattern $\mathbf{F}$ is non uniform and unknown. After a simple but cumbersome computation we obtain:

$$\mathcal{R}_{Funknown} = \frac{1}{4\sigma^2}\left\{\sqrt{D_a^2+W_a^2}+\sqrt{D_b^2+W_b^2}-\sqrt{(D_a+D_b)^2+(W_a+W_b)^2}\right\}, \quad (2)$$

where $D_v = \sum_{i\in\Omega_v}(X_i^2 - Y_i^2)$ and $W_v = 2\sum_{i\in\Omega_v} X_iY_i$ with $v = a, b$. It can be shown that this detector is only sensitive to OSC contrast.

In order to take into account the intensity contrast, one may use the detector adapted to uniform illumination ($F_i = F_0, \forall i \in [1, N]$) whose expression is easily computed:

$$\mathcal{R}_{uni} = \frac{1}{2\sigma^2}\frac{N_aN_b}{N_a+N_b}\left[\left(\sum_{i\in\Omega_a}\frac{X_i}{N_a}-\sum_{i\in\Omega_b}\frac{X_i}{N_b}\right)^2+\left(\sum_{i\in\Omega_a}\frac{Y_i}{N_a}-\sum_{i\in\Omega_b}\frac{Y_i}{N_b}\right)^2\right]. \quad (3)$$

We applied detectors $\mathcal{R}_{uni}$ and $\mathcal{R}_{Funknown}$ on experimental images of a metalic object on a diffusive background (see Fig.1). We present successively the intensity image, the OSC image, the results of $\mathcal{R}_{uni}$ and $\mathcal{R}_{Funknown}$ detection. This series of images are realized for three different values of correlation length of the spatial variations. The correlation length l_c, is defined as the half width at half maximum (HWHM) of the autocorrelation function of the illumination pattern.

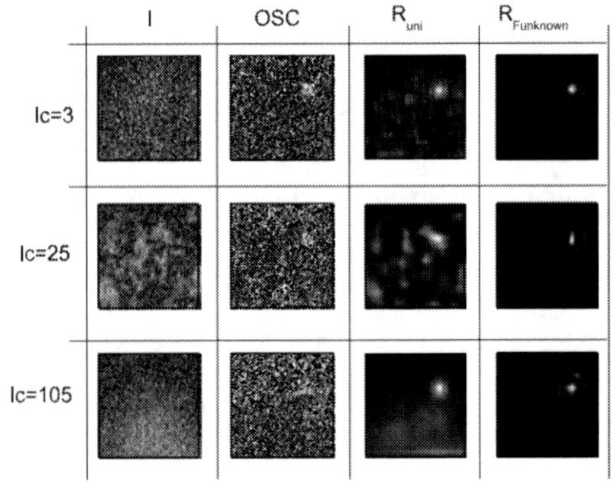

Figure 1. Detection in scenario corresponding to OSC contrast only. Horizontally: intensity image, OSCI, the results of $\mathcal{R}_{uni}$ and $\mathcal{R}_{Funknown}$ detection. Vertically: those images are realized with illumination pattern with different coherence length: $l_c = 3$, $l_c = 25$ and $l_c = 105$.

We observe that, on the one hand, detector $\mathcal{R}_{Funknown}$ shows really good performance and is able to detect the target under all different illuminations. On the other hand, the detector $\mathcal{R}_{uni}$ is less efficient than $\mathcal{R}_{Funknown}$ for all l_c and is significantly affected by the type of illumination. Let us linger on the case $l_c = 25$, which correspond to an illumination pattern whose correlation length approximatively matches the size of the target. In this case we potentially observe $\mathbf{F_a} \neq \mathbf{F_b}$, which leads to high probability of false alarm. The detector actually consider the variations of the illumination as a contrast in intensity. When $l_c = 3$ or $l_c = 105$, the illumination variations are either smaller or larger than the size of the targets which implies $\mathbf{F_a} \approx \mathbf{F_b}$. This explains why, in these cases, $\mathcal{R}_{uni}$ has better performances than when $l_c = 25$.

To conclude, we demonstrated that the spatial distribution of the illumination, and the existing contrats in the scene must be taken into account to decide between both detectors. An interesting perspective will consist in integrating, in the model, the photon noise, and to generalize the approach to other image processing tasks such as segmentation for instance.

References

[1] Goudail F and Réfrégier P 2001 *J. Opt. Soc. Amer. A* **18** 3049–3060
[2] Bénière A, Goudail F, Alouini M and Dolfi D 2009 *JOSA A* **26** 1678–1686

Miniaturized remission sensor for carbon dioxide detection

T Martan[1] and M Will[2]

[1] Institute of Photonics and Electronics, AS CR v.v.i. Prague, Chaberska 57, 18251 Czech Republic
[2] CiS Forschungsinstitut für Mikrosensorik und Photovoltaik GmbH, Konrad-Zuse-Strasse 14, D-99099 Erfurt, Germany

E-mail: martan@ufe.cz

Abstract. Recently, optical sensors for detection of carbon dioxide (CO_2) have been explored for variety of applications in chemistry, industry, and medicine. This paper deals with the development of a planar optical remission sensor employing a dye immobilized in a polymer layer designed for gaseous CO_2 detection. The principle of CO_2 detection was based on colour changes of Tetraethylammonium Cresol red immobilized in a special composed polymer layer that was irradiated by LED diodes. Absorption properties of the dye were changed due to its chemical reaction with CO_2 and corresponding colour changes were detected by PIN diodes. These changes were analyzed by using a PC-controlled board connected by USB. The sensitivity, response time, and the detection limit of the remission sensor were characterized.

1. Introduction

The detection of gaseous carbon dioxide (CO_2) is important in medicine, chemical or bio-chemical processes as well as in many industrial processes [1]. Recently, planar or fibre-optic approaches have been used for CO_2 sensing [2, 3].

This paper deals with the development of a miniaturized planar optical remission sensor intended for concentration measurements of CO_2 in gaseous mixtures. The sensor is based on a Micro–Optical REmission Sensor (MORES) platform that was modified with a sensitive polymer layer based on a dye immobilized in a special-composed polymer [4]. The absorption spectrum of the dye is significantly changed during its chemical reaction with CO_2. It means that the dye allows us to convert specific chemical information on CO_2 presence and concentration into light-absorption changes which were further processed by a PC. A prototype of a sensor system has been built by using the MORES sensor platform with the sensitive layer, so-called evaluation board, and PC controlling programmes.

2. Optical sensor system

The remission sensor system was designed as a robust sensing tool containing the MORES sensor chip provided with a removable layered detection structure. This structure was based on a dye immobilized in a polymer film that was covered by porous protective Teflon foil. The sensitive layer was spin-coated by using a cocktail containing Tetraoctylammonium Cresol red dye, Tetraoctylammonium hydroxide, Tetrahydrofuran, and plasticizer-free polymer TT-1085A on a glass substrate. Scanning White Light Interpherometric microscope (Zygo NewView 2000) was used for the characterization of the spin-coated layer. A thickness of the layer was of 900 nm when using a spin-coating velocity of 1000 [1/min]. Advantages of this sensitive layer consist in colour reversibility related to the reversibility of the chemical reaction of the dye with CO_2 without leaching or bleaching, good spectral and mechanical properties. During concentration changes of CO_2 in a stream with nitrogen it was observed that the absorbance increases in a wavelength range to about 440 nm due to protonation and simultaneously that the absorbance significantly decreases in a range from 440 to 590 nm. It means that the colour of the sensitive layer was changed from blue to yellow with increased CO_2 concentrations.

The sensitive layer was tested by using a spectral analyzer (Prema SPM 9001). The sensitive layer spin-coated on a glass substrate was closed in a measuring chamber and the concentration of CO_2 in a stream with nitrogen was precisely changed. Absorption changes of the sensitive layer were evaluated for concentration changes in a range from of 0 to 100000 ppm. Nearly linear calibration curve was observed in the range of CO_2 concentrations up to 10000 ppm (1% vol.). The nonlinearity of the calibration curve above 1% vol. was observed which can be explained by reaching the CO_2 saturation in the layer. Time responses corresponded to the thickness of the sensitive layer. The response of about 1s was determined for a layer thickness of 900 nm. The best sensitivity was of $2.22.10^{-4}$ 1/ppm and the limit of detection of the sensor was of 0.02 ppm. The MORES sensor chip was developed for the purpose of the evaluation of colour changes of sensitive polymer layer. The sensor chip was designed making possible to integrate three LED diodes in a central cavity (200 μm deep) and two PIN diodes symmetrically situated around this cavity. Emission wavelengths of the light emitting diodes (450 nm, 510 nm, and 580 nm) were optimized for colour changes of the dye. The dimensions of the MORES sensor chip were of 8 x 8 x 1.5 mm^3 and dimensions of each LED diode were of 300 x 300 μm^2. Light changes response of the irradiated polymer film was detected by the PIN diodes with dimensions 2 x 7.8 mm^2. An integrated temperature sensor and monitor diodes for corrections of measured data were attached into the body of the MORES sensor module. Measured data were pre-processed in an evaluation board that was assembled on basis of a Texas Instrument controller MSC1211Y5. This evaluation board is able to mediate communication between a PC and the MORES sensor module via USB port. Information about CO_2 concentration was processed and displayed by the PC.

3. Conclusion

A rugged portable planar remission sensor based on a sensitive polymer layer containing Tetraoctylammonium Cresol red dye, LED and PIN diodes was designed and developed. The functionality of the sensor that was provided with the sensitive layer prepared by spin-coating was successfully tested. The limit of detection of the sensor of 0.02 ppm was found. The best sensitivity of $2.22.10^{-4}$ [1/ppm].was reached. A nearly linear calibration curve was observed in a concentration range from 0.02 to 10000 ppm. The nonlinearity above the upper limit can be explained by CO_2 saturation in the dyed-polymer layer. The shortest response of 1s was observed for a layer thickness of 900 nm.

4. Acknowledgement

This research has been supported by the NASCENT (Nanomaterials for Application in Sensors, Catalysis and Emerging Technologies) research training network No. MRTN-CT-2006-033873 and by the Institute of Photonics and Electronics, AS CR research project No. AV0Z20670512.

References

[1] Hirschfeld T, Miller F, Thomas S, Miller H, Milanovich F, Gaver R 1987 *J. Light. Tech.* **5** 1027
[2] Lang Th, Wiemhöfer H-D, Göpel W 1996 *Sens. and Actuat.* B **34** 383
[3] Petersen J I, Vurek G G 1984 *Science* **224** 123
[4] Will M, Martan T, Müller R, Brodersen O, Mohr G J 2008 *Proc. SPIE* **0277-786X** 7138 0A-1

Tapered optical fibres for local pH detection

T Martan, M Pospisilova, J Aubrecht, J Mrazek, O Podrazky, I Kasik, and V Matejec

Institute of Photonics and Electronics, AS CR v.v.i. Prague, Chaberska 57, 18251 Czech Republic

E-mail: martan@ufe.cz

Abstract. This paper deals with preparation of optical probes based on tapered optical fibres and their using for local detection of pH in samples simulating native conditions of plant cells. Optical probes, so-called V-tapers were prepared for this purpose. Fluorescence pH transducer 2',7'-Bis(2-carbonylethyl)-5(6)-carboxyfluorescein was immobilized onto the end-face of cut V-tapers in a thin xerogel layer. Changes of fluorescence-intensity spectra caused by pH changes were employed for pH measurements. Indium-tin oxide or aluminium coated fibre probes with suitable optical and mechanical properties were successfully tested. It has been confirmed that the selected pH transducer is suitable for pH detection from 5.0 to 7.0. This approach has allowed us to determine extra-cellular pH of in-vitro samples.

1. Introduction

Local detection of intra- and extra-cellular pH has become interesting issue e.g. for botanists, physiologists and also for researchers dealing with optical sensing [1-5]. Tapered optical fibres can be employed as appropriate tools for investigating chemical changes in plant cells, e.g. tobacco leafs with dimensions up to approx. 50x10 μm. We have developed fibre-optic probes based on tapered optical fibres with diameters and properties of the tips that can be used for measurement of intra- or extra-cellular pH of plant cells [6]. These fibre-probes have been intended especially for monitoring of auxins (phytohormones essential for plant growth), which are week acids, pH 5.0 – 7.0 [7, 8]. The prepared probes with high spatial resolution employ an appropriate optical-chemical transducer for the given pH scale. These probes have suitable optical and mechanical properties making possible the penetration of the tip through surfaces (cell walls and membranes) of plant cells.

2. Tapered optical fibres for pH detection

2.1. Tapered optical fibres

Fibre tapers were manufactured from polymer-clad silica (PCS) fibres (125/200 μm) by using a PC-controlled tapering machine. Tapers in so-called "V-taper" configuration with waist diameters from 60 μm down to 4 μm were prepared. V-tapers were made by fusing two fibre tapers together. The surfaces of the tapers were sputtered with a layer of Indium-tin oxide (ITO) of a thickness ~ 150 nm or aluminium of a thickness ~ 100 nm immediately after the tapering. Optical-chemical pH transducer 2',7'-Bis(2-carbonylethyl)-5(6)-carboxyfluorescein (BCECF) was immobilized onto the end-face of the taper tip by using a thin xerogel layer prepared from a tetraexthoxysilane (TEOS) sol in which the transducer was dissolved [8].

2.2. pH detection

Fluorescence spectroscopy was employed for pH measurements in buffers with a pH range from 5.0 to 7.0 as well as in plant cell cultures simulating native conditions. The BCECF transducer was excited by a laser diode (LD) beam at wavelength of 473 nm (FWHM = 11 nm). Light from LD source was coupled into one PCS pigtail of the V-taper and spectra of the fluorescence intensity were directly collected from second PCS pigtail of V-taper by a spectrometer. Drops of pH buffers (a volume of 1 ml) were used for the calibration of the probes. Maxima of fluorescence responses to pH 5.0 and 6.2 were observed at wavelength of 580 nm. Measured curves of excitation and fluorescence response were completely separated without any overlaps. A shift of fluorescence maximum of about 107 nm from wavelength of excitation maximum was observed. This shift can be attributed to the influence of xerogel matrix encapsulating the transducer. The response to a particular pH value depended on time. This phenomenon can be caused by saturation of the porous xerogel layer.

3. Conclusion

Fluorescence probes based on aluminium or ITO-coated "V-tapers" with immobilized BCECF transducer on the fibre tip have been prepared. The developed probes have suitable optical-chemical and mechanical properties. A novel experimental set-up based on excitation and detection with V-taper probe has successfully been implemented. A shift of fluorescence maximum of 107 nm from wavelength of excitation maximum was observed. This significant shift made possible to detect fluorescence response without optical filtering of the excitation. This approach has allowed us to determine extra-cellular pH of in-vitro samples in a range from pH 5.0 to pH 7.0.

4. Acknowledgement

This research has been supported by the Czech Science Foundation contracts No. 102/08/P639.

References

[1] Paciorek T, Zazimalova E, Ruthardt N, Petrasek J, Stierhof Y D, Kleine-Vehn J, Morris D A, Emans N, Jurgens G, Geldner N, Frim J 2005 *Nature* **435** 1251

[2] Rosenzweig Z, Kopelman R 1995 *Anal. Chem.* **67** 2650

[3] Tan W, Shi Z, Kopelmann R 1992 *Anal. Chem.* **64**, 2985

[4] Vo-Dinh T, Kasili P 2005 *Anal. Bioanal. Chem.* **382** 918

[5] Rosenzweig Z, Kopelman R 1996 *Anal. Chem.* **68** 1408

[6] Martan T, Kasik I, Podrazky O, Mrazek J, Pospisilova M, Aubrecht J, Matejec V, Kanka J 2008 *Proc. MADICA* **6** A53-1

[7] Podrazky O, Mrazek J, Seidl M, Kasik I, Tobiska P, Matejec V, Martan T, Aubrecht J 2007 *Proc. SPIE* **6585** 65850Y.1

[8] Martan T, Podrazky O, Kasik I, Pospisilova M, Matejec V, Kanka J 2008 *Proc. Europtrode* **IX**, 276

Dynamic wavefront sensing and correction with low-cost twisted nematic spatial light modulators

Vicente Durán[1], Vicent Climent[1], Jesus Lancis[1], Enrique Tajahuerce[1], Salvador Bará[2], Justo Arines[3], Jorge Ares[3], Pedro Andrés[4], and Zbigniew Jaroszewicz[5]

[1]GROC•UJI, Physics Department-INIT, Universitat Jaume I, E12080 Castelló, Spain

[2]Departamento de Física Aplicada, Universidade de Santiago de Compostela, 15782 Compostela, Galiza, Spain

[3]Departamento de Física Aplicada, Universidad de Zaragoza, Zaragoza, Spain

[4]Departamento de Óptica, Universitat de València, E46100 Burjassot, Spain

[5]Institute of Applied Optics, Kamionkowska 18, 03-805 Warsaw, Poland, and National Institute of Telecommunications, Szachowa 1, 04-894 Warsaw, Poland

E-mail: lancis@fca.uji.es

Abstract. Off-the-shelf twisted nematic liquid crystal displays (TNLCDs) show some interesting features such as high spatial resolution, easy handling, wide availability, and low cost. We describe a compact adaptive optical system using just one TNLCD to measure and compensate optical aberrations. The current system operates at a frame rate of the order of 10 Hz with a four level codification scheme. Wavefront estimation is performed through conventional Hartmann-Shack sensing architecture. The system has proved to work properly with a maximum rms aberration of 0.76 microns and wavefront gradient of 50 rad/mm at a wavelength of 514 nm. These values correspond to typical aberrations found in human eyes. The key of our approach is careful characterization and optimization of the TNLCD for phase-only modulation. For this purpose, we exploit the so-called retarder-rotator approach for twisted nematic liquid crystal cells. The optimization process has been successfully applied to SLMs working either in transmissive or in reflective mode, even when light depolarization effects are observed.

1. Introduction

Wavefront sensing and compensation is today an enabling technology for biomedical optics. There is correspondingly a trend towards the development of reliable, low-cost, easy-to-use devices for their transfer to clinical practice. General-purpose spatial light modulators (SLMs), like TNLCDs used in video projection devices, offer some interesting features for this task. They can be configured to act basically as segmented (pixelated) piston correctors, offering a relatively high spatial resolution (of the order of 0.5 to 0.7 megapixels for inch-sized displays) and easy addressability and control. They are also low-cost devices widely available, and it can be anticipated that their performance will improve in the near future, driven by the needs of the consumer electronics industry. They share with the deformable mirrors their ability to compensate dynamic eye aberrations (although with a lower temporal band-width), and share with the phase plates their high spatial resolution [1].

2. Measurement and compensation of optical aberrations using a single spatial light modulator

Figure 1 shows the basic block diagram of the adaptive system developed in this work [2]. The core of the system is a TNLCD (SLM) located in the main path of an aberrated beam A. During the measurement step an array of diffractive microlenses is encoded in the TNLCD to implement a Hartmann-Shack wavefront sensor. An image of its focal spots is taken at the CCD with the aid of a beam splitter (BS) and the beam aberration is determined with the aid of a conventional computer (PC). This aberration, with opposite sign, is subsequently encoded into the TNLCD replacing the microlens array, starting in this way the compensation step. RO1 and RO2 are two (optional) relay optics modules, useful to project onto the TNLCD the input pupil of the system, and to project the TNLCD onto the exit pupil; besides, they allow for matching the respective pupil sizes, making an optimum use of the available spatial resolution of the TNLCD. During the measurement step the exit beam at B is distorted due to its interaction with the microlens array and this implementation of the adaptive unit provides no useful beam to the optical systems located behind it. However, the measurement time can be kept within very small values using an efficient design of the image acquisition and processing routines. Once the aberration of the incoming beam is compensated, no further measurement has to be made until the quality of the output beam is degraded above a user-defined threshold.

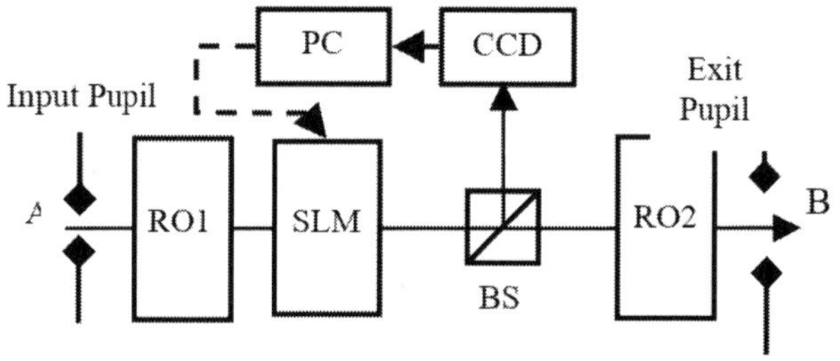

Figure 1. Block diagram of the single SLM adaptive setup (see text for details).

Figure 2 shows the CCD images obtained at three consecutive slots of time in a typical demonstration run. Fig. 2 (left), taken during the measurement step, shows at the left side the microlens array foci, and at the lower right the image of the LED source given by the system, strongly distorted due to the pass of the aberrated beam trough the microlens array; Fig. 2 (center) shows an enlarged view of the original aberrated LED image, with the TNLCD in off state. Finally, Fig. 2 (right) shows the corrected image, after the TNLCD was fed with the compensation pattern. The collapsing of the aberrated PSF after compensation can be easily noticed.

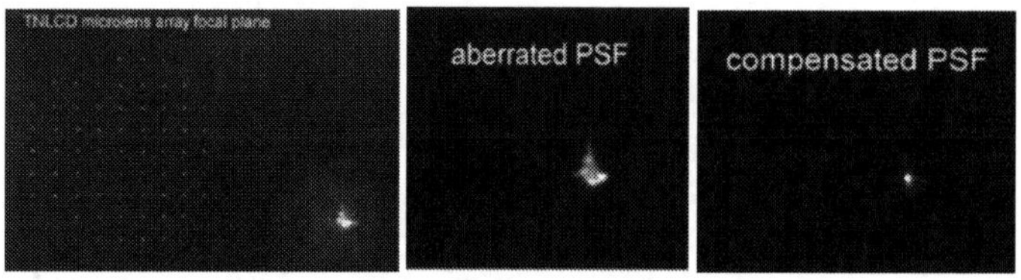

Figure 2. CCD images obtained at three consecutive time intervals in a typical demonstration run.

References

[1] Durán V, Climent V, Tajahuerce E, Jaroszewicz Z, Arines J, Bará S 2007 *J. Biomed. Opt.* **12**, 014037

[2] Arines J, Durán V, Jaroszewicz Z, Ares J, Tajahuerce E, Prado P, Lancis J, Bará S, Climent V 2007 *Opt. Express* **15**, 15287-15292

Journal of Physics: Conference Series **206** (2010) 012019 doi:10.1088/1742-6596/206/1/012019

Optically Compressed Sensing by under sampling the polar Fourier plane

A Stern[1], O Levi[2], and Y Rivenson[3]
[1] Electro Optical Department, Ben Gurion University of the Negev, Beer-Sheva 84105, Israel
[2] Department of industrial engineering, Ben Gurion University of the Negev, Beer-Sheva 84105, Israel
[3] Department of Electrical and Computer Engineering, Ben-Gurion University of the Negev, Beer-Sheva 84105, Israel

E-mail: [1]`stern@bgu.ac.il`, [2]`levio@bgu.ac.il`, [3]`rivenson@ee.bgu.ac.il`

Abstract. In a previous work we presented a compressed imaging approach that uses a row of rotating sensors to capture indirectly polar strips of the Fourier transform of the image. Here we present further developments of this technique and present new results. The advantages of our technique, compared to other optically compressed imaging techniques, is that its optical implementation is relatively easy, it does not require complicate calibrations and that it can be implemented in near-real time.

Compressed sensing (CS) [1] – or compressive sensing as it is also known – is a new field that has been rapidly emerging over the past few years. CS suggests a new framework for *simultaneous* sampling and compression of signals. As opposed to the presently used framework in which as much data as possible is first collected and then the redundant data is discarded by digital compression techniques, CS seeks to minimize the collection of redundant data in the acquisition step. Despite the young age of the CS theory, several implementation of CS concept in imaging have already appeared [3]. In this work we further elaborate a compressive imaging (CI) approach introduced in [2] which captures image projections containing information of radial strips of the Fourier domain of the image. This CI approach can be implemented with coherent and incoherent imaging with a vector (row) sensor that scans the field-of-view by rotational motion. As illustrated in Fig.1 the approach in [2] takes benefit of very good trade between implementation complexity and acquisition time. By complexity we generally refer to system implementation complexity, system calibration, reconstruction complexity and other computational and storage requirements [4].

One way to capture radial strips in the Fourier domain, applicable for incoherent imaging, is by utilizing a cylindrical lens as described schematically in [2]. The system performs an integral projection and such that the sensor measures the Radon transform of the image [2].

Figure 2 presents image reconstruction from data captured with an optical experiment based on the scheme on the scheme described in Fig. 4 in [2]. The system uses a cylindrical lens to capture the Radon transform of the image with a linear sensor. Therefore, according to the central slice theorem [2] the captured field is related to the 2D Fourier transform of the image on a radial strip. The object is an image of a goat imaged with the incoherent CI system working in the visible spectrum range. In this experiment, the vector sensor had 200 effective pixels and the number of rotated projections was 36. The rotational scan covered 180 degrees. Note that

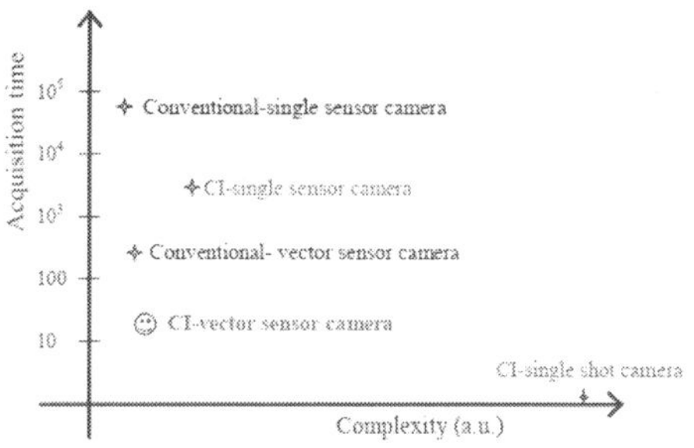

Figure 1. Heuristic diagram of acquisition time versus complexity (a.u.) of conventional and CI systems for an image of 256x256 pixels

the reconstruction presented in Fig.2 is obtained from only 7200 measurements, compared to 40,000 measurements required for a conventional imaging system to obtain an image of 200x200 pixels. That is, a compression ratio of 5.55:1 is obtained optically with negligible lost of visual information. For illustration of the ability of the method to preserve the visual information, Fig.2(b) presents the image that would be obtained using a conventional imaging system that scans horizontally the same field of view, using the same sensor, and capturing the same number (6400) of samples.

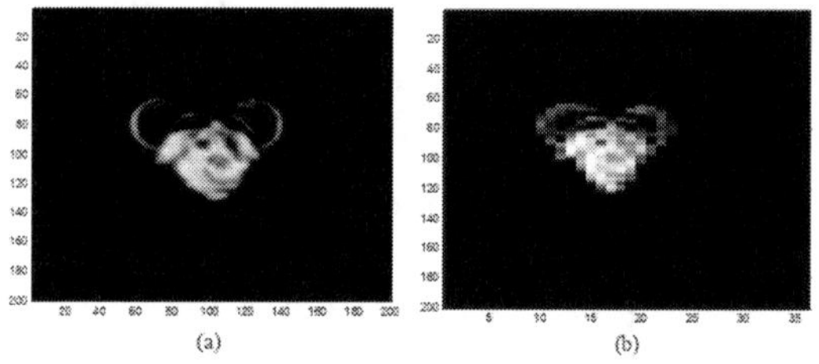

Figure 2. (a) Image reconstructed from rotational scanning with the incoherent CI system in the visible; a vector sensor having 200 pixels performed rotational scan of half a circle and took 36 exposures. (b) image that would be obtained with a conventional scanning system scanning horizontally the field of view and capturing the same number of samples as in (a).

The CI technique considered here exhibits many attractive features compared to other CI techniques (see Fig.1). Let us consider the acquisition time of an $n \times n$ pixels image for the various imaging techniques. A single sensor conventional camera takes $n \times n$ exposures to capture the image. A single sensor CI camera needs c times less exposures for capturing the same image, were c is typically between 5-10. Single shot CI cameras require only one exposure; however, this comes on account of complexity and computational burden, while the proposed sensor requires no calibration or storage of "system matrix" [4]. Thus, we might conclude that the CI technique described in this work technique exhibits a very good trade between acquisition time and system complexity.

References

[1] `http://www.dsp.ece.rice.edu/cs/`
[2] Stern A 2007 *Optics Letters* **32** 3077
[3] `http://nuit-blanche.blogspot.com/2008/06/cs-compressive-imaging.html`
[4] Rivenson Y and Stern A 2009 Proc. 16th Int. Conf. on Digital Signal Processing (Santorini:IEEE) p 001

Constructive role of sensors nonlinearities in the acquisition of partially polarized speckle images

Agnès Delahaies[1], David Rousseau [1],
Denis Gindre[2], François Chapeau-Blondeau[1]

(1) Laboratoire d'Ingénierie des Systèmes Automatisés (LISA),
Université d'Angers, 62 avenue Notre Dame du Lac, 49000 Angers, France.
(2) Laboratoire des Propriétés Optiques des Matériaux et Applications (POMA),
Université d'Angers, 2 boulevard Lavoisier, 49000 Angers, France.

E-mail: `david.rousseau@univ-angers.fr`

Abstract. We study the impact of the level of the speckle noise on data acquisition in a partially polarized coherent imaging system with the presence of a nonlinearity in the imaging sensor characteristic. In perfectly linear acquisition conditions, due to the essentially multiplicative action of the speckle, the image contrast is unchanged as the speckle noise level increases, and so it has no impact on the quality of the acquired images. On the contrary, in nonlinear conditions the acquisition is affected by the speckle noise level. However, this effect of the speckle is not always detrimental. We show that, in definite nonlinear conditions, there is usually an optimal level of the speckle noise that leads to a maximum quality of the acquired images. We theoretically analyze such nonlinear regimes with partially polarized speckled images. We specifically exhibit the existence of an optimal speckle noise level in the interesting case of images realized only by a depolarization contrast. Illustrations are given with a simple 1-bit hard limiter and binary images. Then, we propose and discuss as perspectives an experimental optical setup to confront theory and experiment.

The present study is connected to the field of stochastic resonance and improvement by noise in nonlinear information processing, and it considers possible extensions and applications to optical signals and imaging. In this framework, we have recently [2, 3, 5, 4] demonstrated the possibility of coherent imaging situations where nonlinearities (threshold [2, 3], saturation [5]) in the sensor response, cooperating with noise, can give way to an improvement of the acquired image in comparison with the image produced by a perfectly linear sensor. The images considered in [2, 3, 5, 4] are strongly affected by speckle noise and characterized by their statistical properties. The image acquisition is shown to be improved when the speckle noise is tuned up to a level which is large enough to force the sensor to operate in the nonlinear parts of its response. Such possibilities of exploiting the speckle noise in order to draw benefit from sensor nonlinearities are illustrated with images of calibrated transparencies with multiple [2] or binary [3, 5, 4] distributed gray levels. The model of speckle in [2, 3, 5, 4] was a simple model of fully developed and polarized speckle. Here, we propose to extend the work from [2, 3, 5, 4] to partially polarized speckle noise [1, 9].

Results are illustrated in Fig. 1 where the input image is a binary image constituted by two regions occupying identical surfaces and presenting a contrast of optical properties in terms of transparency and depolarization. In Fig. 1, the input image is then corrupted by a multiplicative speckle noise and the output image is delivered by a sensor with input–output characteristic taken as a hard limiter with threshold equal to 1.Fig. 1 gives an illustration of the impact of the speckle noise level for various contrast in transparency and depolarization on the input–output image transmission assessed by a similarity measure chosen as the input–output mean square error. We observe in Fig. 1 a nonmonotonic evolution of the mean square error which can be improved when the level of the speckle noise increases. As visible in Fig. 1, the input–ouput mean square error reaches its minimum value when the level of the speckle noise is tuned at an optimal nonzero value. A specifically interesting feature in Fig. 1 is that the nonmonotonic evolution of the mean square error as a function of the speckle amplitude is preserved even for images with a contrast in polarization and no contrast in transparency (case of $I_1 = I_0 = 1$).

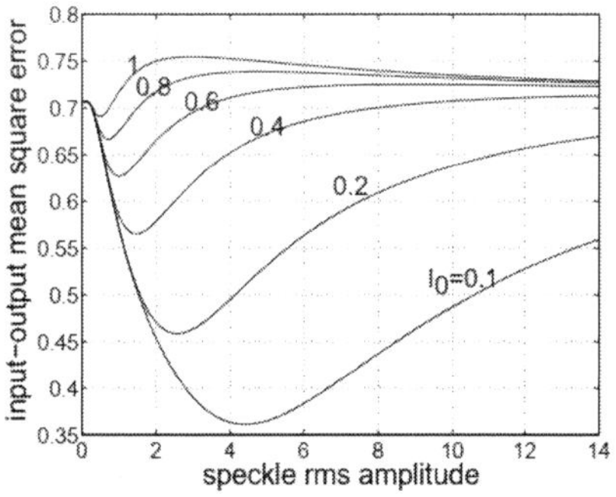

Figure 1. Mean square error as a function of the speckle noise rms amplitude for various transparencies I_0 and $I_1 = 1$ in a binary input image.

As a perspective, it would now be interesting to confront, as done in [3, 5], experiment and theoretical modeling. It would be an easy task to build a binary input image with a maximal contrast in depolarization properties. For example [1], one can think of a scene constituted with a rough dielectric region such as paper typically producing a reflected wave with a polarization degree ≈ 0 and a rough metallic region to have a polarization degree ≈ 1. Yet, it would be more difficult to find a full set of surfaces presenting depolarization properties capable of reflecting waves with a degree of polarization uniformly covering the whole range $[0, 1]$. Alternatively, a partial depolarization of an incident polarized wave can also be controlled by liquid crystal devices (like in [6, 7]). And also, multimode optical fibers are known [8] to partially depolarize light. Because they are cheaper than liquid crystal devices and easier to implement than a set of surfaces of variable roughness, a set of variable length optical fibers appears as a good candidate to propose a future experimental validation of the theoretical results presented here.

References

[1] J. W. Goodman, *Speckle Phenomena in Optics*, (Roberts & Company, Englewood, 2007).
[2] S. Blanchard, D. Rousseau, D. Gindre, and F. Chapeau-Blondeau, Noise-assisted image transmission with speckle noise, *Proceedings 5th International Conference on Physics in Signal and Image Processing*, Mulhouse, France, 31 Jan. - 2 Feb. 2007.
[3] S. Blanchard, D. Rousseau, D. Gindre, and F. Chapeau-Blondeau, Constructive action of the speckle noise in a coherent imaging system, *Optics Letters* **32**, 1983–1985 (2007).
[4] S. Blanchard, D. Rousseau, D. Gindre, and F. Chapeau-Blondeau, Benefits from a speckle noise family on a coherent imaging transmission, *Optics Communications* **281**, 473–479 (2008).
[5] F. Chapeau, D. Rousseau, S. Blanchard and D. Gindre, Optimizing the speckle noise for maximum efficacy of data acquisition in coherent imaging, *Journal of the Optical Society A* **25**, 1287–1292 (2008).
[6] N. Diorio, M. Fisch, J. West, Filled liquid crystal depolarizers, *Journal of Applied Physics* **90**, 3675–3678 (2001).
[7] M. Honma, T. Nose, Liquid-crystal depolarizer consisting of randomly aligned hybrid orientation domains, *Applied Optics* **43**, 4667–4671 (2004).
[8] P. Agrawal, *Nonlinear Fiber Optics*, 4th edition (Academic Press, Boston, 2007).
[9] P. Réfrégier, J. Fade, M. Roche, Estimation precision of the degree of polarization from a single speckle intensity image, *Optics Letters* **32**, 739–741 (2007).

Journal of Physics: Conference Series **206** (2010) 012021 doi:10.1088/1742-6596/206/1/012021

General Solution Of Two-Dimensional Beam-Shaping With Two Surfaces

K.–H. Brenner

Chair of Optoelectronics, University of Heidelberg, B6, 23-29,
D 68131 Mannheim, Germany

E-mail: brenner@ziti.uni-heidelberg.de

Abstract. Beam shaping is a technique, by which a known input irradiance is transformed into a desired output irradiance by changing the local propagation vector of the wave front. Unlike one-dimensional beam-shaping, which leads to a simple differential equation which can be integrated in a straight forward manner, the two-dimensional beam shaping problem leads to a Monge-Ampere type equation, which is difficult to solve. In this paper, we generalize the problem to refractive and reflective systems and use to shifted-base-function approach to obtain a general solution.

1. Introduction

Beam shaping is used to transform a given intensity distribution into a different, desired intensity distribution. Applications of one-dimensional (rotationally symmetric) beam shaping are typically in the area of illumination and high power laser operations. Unlike one-dimensional beam-shaping [1], which leads to a simple differential equation which can be integrated in a straight forward manner, the two-dimensional beam shaping problem leads to a nonlinear Monge-Ampere type equation [2,3], for which numerical solutions [4,5] are difficult to obtain.

2. Optical beam shaping with two surfaces

Beam shaping can be realized by diffractive, by reflective or refractive optics. In this paper, we consider refractive optics in a telescopic geometry. The input medium and the output medium are assumed to have the same refractive index, here denoted by n_1. The intermediate medium, which has a thickness of D on the optical axis has a refractive index of n_2. In a notation similar to Shealy [6], the input coordinates are denoted by small letters (x,y) and the output coordinates are denoted by captial letters (X,Y). Any beam shaping problem can be decomposed into a sequence of two tasks: The mapping of the coordinates (x,y) to the coordinates (X,Y) and the determination of the optical surface S, required to achieve this task. The mapping determines the brightness $I_1(X,Y)$ in the exit plane as function of the brightness in the entrance plane $I_0(x,y)$ according to the local energy conservation law.

$$I_0\, dA_0 = I_1\, dA_1 \tag{1}$$

The brightness in the entrance plane $I_0(x,y)$ is typically defined by the light source. Introducing a surface S modelled by a shifted base function approach [7], the energy law leads to a Monge-Ampere type differential equation

$$I_1\left(x+S_x, y+S_y\right)\cdot\left[\left(1+S_{xx}\right)\left(1+S_{yy}\right)-S_{xy}^2\right]-I_0\left(x,y\right)=0 \tag{2}$$

where we have abbreviated the first order partial derivatives by S_x, S_y and the second order partial derivatives by S_{xx}, S_{xy}, S_{yy}. By applying a special minimization technique, the surface can be derived.

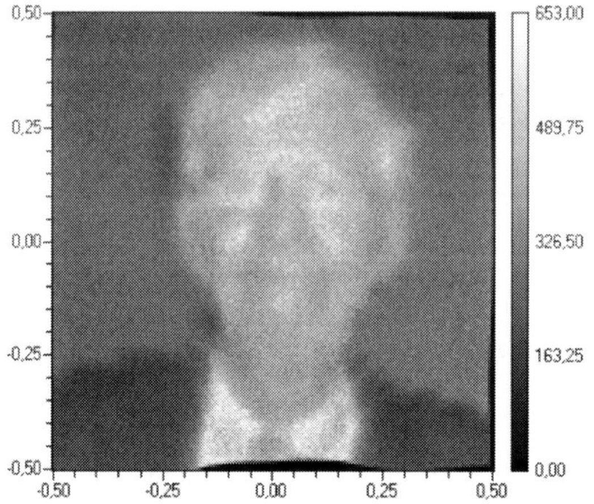

Figure 1. Intensity distribution obtained by Monte-Carlo ray trace.

3. Examples

In the following we consider a gray valued pattern as an example. With 32x32 base points chosen for the SBF-approximation, the gray level picture in fig. 1, a photograph of A. Lohmann shows the Monte-Carlo ray trace intensity obtained from this surface. The calculation time for this example was in the order of 5 minutes.

4. Conclusion

We established a relationship between the coordinate mapping and the gradient of the surfaces. By minimizing a differential equation, the optimum surface for generating a given intensity distribution can be found. The solution can be applied to both, reflective and refractive beam shaping with two surfaces. As model function for the surfaces, we used with the shifted-base-function (SBF) approach.

References

[1] F. Dickey and S. Holswade, Laser Beam Shaping: Theory and Techniques (Marcel Dekker, 2000)

[2] E. J. Dean, R. Glowinski, Numerical Analysis Numerical solution of the two-dimensional elliptic Monge–Ampère equation with Dirichlet boundary conditions: an augmented Lagrangian approach, C. R. Acad. Sci. Paris, Ser. I 336 (2003) 779–784

[3] V.I. Oliker, L.D. Prussner, On the numerical solution of the equation $z_{xx}\ z_{yy} - z^2_{xy} = f$ and its discretization. I, Numer. Math. 54 (1988) 271–293

[4] T. Glimm and V. Oliker, "Optical design of two-reflector systems, the Monge–Kantorovich mass transfer problem and Fermat's principle," Indiana Univ. Math. J. 53, 1255–1278 (2004).

[5] V. Oliker, "Optical design of two-mirror beam-shaping systems. Convex and non-convex solutions for symmetric and non-symmetric data," Proc. SPIE 5876, 203–214 (2005).

[6] D. Shealy and S. Chao, "Geometric optics-based design of laser beam shapers," Opt. Eng. 42, (2003) 1–16

[7] K.-H. Brenner, "Shifted Base Functions: An Efficient and Versatile New Tool in Optics", Journal of Physics: Conference Series 139 012002, 11 pp, Workshop on Information Optics (WIO'08), 01.-05.06.08 in Annecy (2008)

Fusion of multistatic synthetic aperture radar data to obtain a superresolution image

Ali Mohammad-Djafari[1], Sha Zhu[1], Franck Daout[2] and Philippe Fargette[3]

[1] Laboratoire de signaux et systmes (L2S), UMR 8506 (CNRS-SUPELEC-Univ Paris Sud 11), 91192 Gif-sur-Yvette, France

[2] SATIE, ENS Cachan, Université Paris 10, 94235 Cachan, France

[3] DEMR, ONERA, 91120 Palaiseau, France

E-mail: djafari,zhu@lss.supelec.fr,fdaout@u-paris10.fr,philippe.fargette@onera.fr

Abstract. When using a single emitter and a single receiver, the Synthetic Aperture Radar (SAR) data gives information in the Fourier domain of the scene over a line segment whose width is related to the bandwidth of the emitted signal. The mathematical problem of image reconstruction in SAR then becomes a Fourier Synthesis (FS) inverse problem. When there are more than one emitter or receiver looking at the same scene, the problem becomes fusion and inversion. In this paper we report a Bayesian joint data fusion and inversion method to obtain a super resolution image. The proposed method shows a good performance on real data obtained at ONERA in France.

1. Introduction to SAR imaging

Synthetic Aperture Radar(SAR) has been widely used for remote sensing, geometry detection, and target recognition. Considering the configuration of SAR, the simplest case is "monostatic", when emitter and receiver are in the same position. In this case, SAR gives information in the Fourier domain of the scene on a set of straight lines. The mathematical problem of image reconstruction becomes an inverse problem of Fourier Synthesis (FS) which is an ill-posed problem. When the emitter and receiver are located differently, as so called "bistatic case", or if there are more than one receiver or emitter, as so called "multistatic case", the problem becomes data fusion and inversion. In this paper we employ the Bayesian estimation approach to solve these inverse problems, particularly we are focusing on multi-static case.

2. Forward modeling

In 2D case, after some mathematical simplification, we can obtain a Fourier Transform relation between the spectrum of the observed signals and the 2D spatial F-T of the scene which can be written as:

$$G(k_x, k_y) = M(k_x, k_y)F(k_x, k_y)$$

where $F(k_x, k_y)$ is the 2D spatuial F-T of the scene $f(x, y)$, $M(k_x, k_y)$ is related to the chacteristic of radar system, with $M(k_x, k_y) = 1$ on the set of points where data is available and $M(k_x, k_y) = 0$ on those points where data is lacking, and $G(k_x, k_y)$ is obtained from the observed signals.

Journal of Physics: Conference Series **206** (2010) 012022 doi:10.1088/1742-6596/206/1/012022

3. Inversion

As we discussed above, the forward problem is going from the target $f(x, y)$ to the data $G(k_x, k_y)$ in the Fourier domain. Oppositely, the inverse problem is to obtain $f(x, y)$ from the data $G(k_x, k_y)$.

3.1. Bayesian estimation approach

The relation between the observed data $G(k_x, k_y)$ and the unknown scene $f(x, y)$ is a linear one, when discretized, we obtain

$$M : \quad \boldsymbol{g} = \boldsymbol{H}\boldsymbol{f} + \boldsymbol{\epsilon}$$

where $\boldsymbol{g}$ is a vector containing all the valus of $G(k_x, k_y)$ on those points where $M(k_x, k_y) = 1$, $\boldsymbol{f}$ is a vector of the pixel value of the discrete scene $f(x, y)$ and $\boldsymbol{H}$ is the DFT matrix and $\boldsymbol{\epsilon}$ represents the errors.

Using this forward model, we can define $p(\boldsymbol{g}|\boldsymbol{f}; \boldsymbol{M})$ and using a prior $p(\boldsymbol{f}|\boldsymbol{M})$, we can get

$$p(\boldsymbol{g}|\boldsymbol{f}; \boldsymbol{M}) = \frac{p(\boldsymbol{g}|\boldsymbol{f}; \boldsymbol{M})p(\boldsymbol{f}|\boldsymbol{M})}{p(\boldsymbol{g}|\boldsymbol{M})}.$$

In this paper we have limited ourselves to the Maximum A Posterior(MAP) estimation which is:

$$\widehat{\boldsymbol{f}} = \arg\max_{\boldsymbol{f}} \{p(\boldsymbol{f}|\boldsymbol{g})\} = \arg\max_{\boldsymbol{f}} \{p(\boldsymbol{g}|\boldsymbol{f})p(\boldsymbol{f})\} = \arg\min_{\boldsymbol{f}} \{-\ln p(\boldsymbol{g}|\boldsymbol{f}) - \ln p(\boldsymbol{f})\}$$

where we choose

$$p(\boldsymbol{g}|\boldsymbol{f}) \propto \exp\{\frac{1}{2\sigma_\epsilon^2}\|\boldsymbol{g} - \boldsymbol{H}\boldsymbol{f}\|^2\}$$

$$p(\boldsymbol{f}) \propto \exp\{\lambda \sum_j |f_j - f_{j-1}|^\beta\} \propto \exp \lambda \sum_j |[\boldsymbol{D}\boldsymbol{f}]_j|^\beta$$

3.2. Multistatic data fusion methods

For multistatic case, the main objective is to develop different methods to obtain superresolution [1] and [2]. Here, we introduce a joint data fusion and inversion method: we use two sets of data $\boldsymbol{G}_1(u, v)$ and $\boldsymbol{G}_2(u, v)$ with their respective relations to $f(u, v)$:

$$\begin{cases} \boldsymbol{g}_1 = \boldsymbol{H}_1\boldsymbol{f} + \boldsymbol{\epsilon}_1 \\ \boldsymbol{g}_2 = \boldsymbol{H}_2\boldsymbol{f} + \boldsymbol{\epsilon}_2 \end{cases}$$

to obtain directly the solution by defining again the MAP estimator:

$$\widehat{\boldsymbol{f}} = \arg\max_{\boldsymbol{f}} \{p(\boldsymbol{f}|\boldsymbol{g}_1, \boldsymbol{g}_2)\} = \arg\min_{\boldsymbol{f}} \{J(\boldsymbol{f})\}$$

where assuming ϵ_1 and ϵ_2 to be Gaussian and independent and the same priors for $\boldsymbol{f}$, we have:

$$J(\boldsymbol{f}) = \frac{1}{\sigma_{\epsilon_1}^2}\|\boldsymbol{g}_1 - \boldsymbol{H}_1\boldsymbol{f}\|^2 + \frac{1}{\sigma_{\epsilon_2}^2}\|\boldsymbol{g}_2 - \boldsymbol{H}_2\boldsymbol{f}\|^2 + \lambda\|\boldsymbol{g}_1 - \boldsymbol{H}_1\boldsymbol{f}\|^\beta$$

4. Experiments on real data

The proposed method has been applied on a few set of real experimental data measured by ONERA. The details of this experimentation are reported in [3]. There are three target compositions (1, 2 and 4 spheres) and for each set of targets, there are two sets of full polar data corresponding to two bandwidth BF1=(1.16-1.98)GHz and BF2=(1.47-1.68)GHz. In the first experiment, we use only one of the bands to do reconstructions and then, we use both bands to do joint data fusion and inversion. Here, we show a few results we obtained using the proposed method compared to a classical Fourier inversion method.

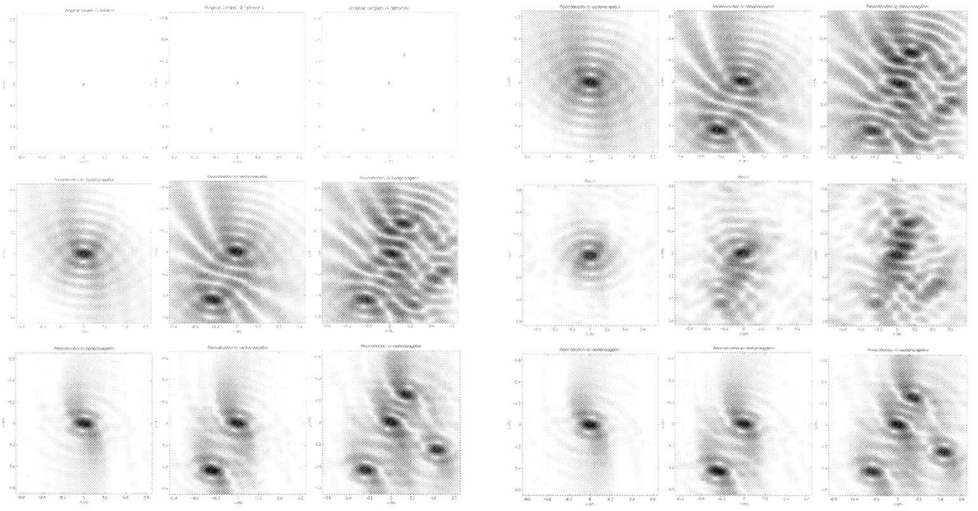

(a) Results on real data (BF1): Orignal targets; Reconstructions by classical FT inversion; Reconstructions by proposed method

(b) Results on real data: Band=BF1; Band=BF2; Fusion of BF1&BF2

Figure 1: Experimental measurement and results

5. Conclusions and perspectives

In this paper, we proposed a Bayesian estimation MAP method to solve the Fourier Synthesis problem in SAR imaging data. The proposed method can be used in all mono- or bi- (multi-) static case. Particularly, in multistatic case, we employed the proposed method to do data fusion and inversion jointly. The experimental results on real data have shown a good performance of our proposed method. The next step, we will focus on two aspects: (a) to use other prior model as Gauss-Markov-Potts prior models [4], [5], [6] and [7]; (b) for real data, we still need to account for polarization information of the measurement.

6. References

[1] Mohammad-Djafari A, Daout F and Fargette P 2009 *ICIP 2009* (Cario, Egypte)
[2] Mohammad-Djafari A, Daout F and Fargette P 2009 *GRETSI 2009* (Dijon, France)
[3] Fargette P Novembre 2008 Compte-rendu d'exprimentation dtc Rapport dga
[4] Humblot F and Mohammad-Djafari A 2006 *EURASIP Journal on Applied Signal Processing* **vol. 2006** 1–16
[5] Mohammd-Djafari A Mar 2008 *the Computer Journal* **52** 126–141
[6] Bali N and Mohammad-Djafari A 2008 *IEEE trans. On Image Processing* **17** 217–225
[7] Féron O and Mohammad-Djafari A 2005 *Journal of Electronic Imaging* **14** 023014

Parallel phase-shifting digital holography based on the fractional Talbot effect

Lluís Martínez-León[1,2], María Araiza-E[3], Bahram Javidi[4], Pedro Andrés[5], Vicent Climent[1,2], Jesús Lancis[1,2] and Enrique Tajahuerce[1,2]

[1] GROC·UJI, Departament de Física, Universitat Jaume I, 12071 Castelló, Spain
[2] INIT (Institute of New Imaging Technologies), Universitat Jaume I, 12071 Castelló, Spain
[3] Laboratorio de Procesamiento Digital de Señales, Universidad Autónoma de Zacatecas, Zacatecas, México
[4] Department of Electrical and Computer Engineering, University of Connecticut, CT 06269-2157, USA
[5] Departament d'Òptica, Universitat de València, 46100 Burjassot, Spain

E-mail: `enrique.tajahuerce@uji.es`

Abstract. A method for recording on-axis single-shot digital holograms based on the self-imaging phenomenon is reported. A simple binary two-dimensional periodic amplitude is used to codify the reference beam in a Mach-Zehnder interferometer, generating a periodic three-step phase distribution with uniform irradiance over the sensor plane by fractional Talbot effect. An image sensor records only one shot of the interference between the light field scattered by the object and the codified parallel reference beam. Images of the object are digitally reconstructed from the digital hologram through the numerical evaluation of the Fresnel diffraction integral. This scheme provides an efficient way to perform dynamic phase-shifting interferometric techniques to determine the amplitude and phase of the object light field. Unlike other parallel phase-shifting techniques, neither complex pixelated polarization devices nor special phase diffractive elements are required. Experimental results confirm the feasibility and flexibility of our method.

1. Introduction

The complex amplitude distribution of a diffracted light beam can be measured and reconstructed through digital holography techniques [1]. Several approaches have been proposed to prevent overlapping of the real and virtual images and the zero order in the reconstruction. One of the most direct methods is recording the digital hologram with off-axis geometries, introducing a tilt between the object and reference beams, but an acceptable trade-off between the image resolution and the size of the reconstructed area must be contemplated. On the other hand, phase-shifting techniques efficiently exploit the spatial resolution of digital sensors. Since several interference patterns are acquired in this scheme, it cannot be used when the object wave front experiences fast changes. Recently, some parallel phase-shifting techniques have been developed. They allow dynamic measurements by recording several phase-shifted interference patterns with just one shot of the image sensor but, usually, pixelated polarization devices or special phase diffractive elements are required.

2. Parallel phase-shifting digital holography based on the Talbot effect

We propose a new method for recording single-shot digital holograms based on the fractional Talbot effect [2]. In our dynamic Talbot-based digital holography system, shown in Fig. 1(a), a binary two-dimensional periodic grating generates a periodic three-step phase distribution with uniform irradiance over the sensor plane by fractional Talbot effect (see Fig. 1(b)). The periodic

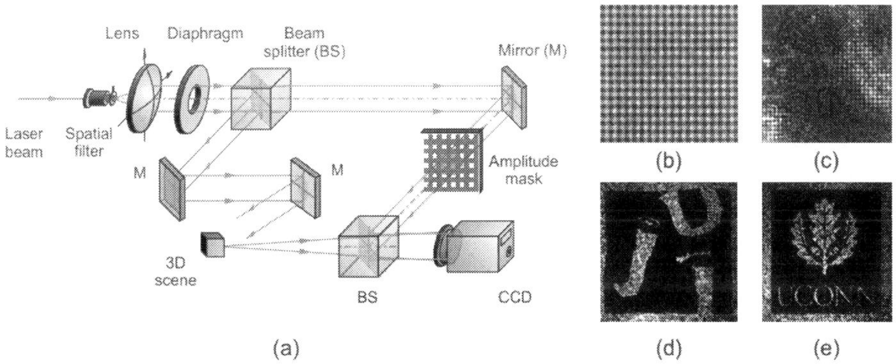

Figure 1. (a) Optical system to perform Talbot digital holography. (b) Gray-level picture of a small region of the phase distribution associated to the fractional Talbot image of the grating, a periodic distribution of three different phases with uniform irradiance. (c) Interference pattern between the object diffraction beam and the reference fractional Talbot image at the CCD sensor. (d) and (e) Reconstruction of the image of two objects located at different distances from the CCD.

distribution codifies the reference beam. The sensor records only one shot of the interference between this reference beam and the light field scattered by an object. In this way, different pixels of the sensor record the interferogram with different phase shifts (see Fig. 1(c)). By sampling and processing the pixelated phase-shifted interferogram we are able to measure the amplitude and phase of the object light field. Therefore we have an efficient way to perform dynamic phase-shifting digital holography. Moreover, neither complex pixelated polarization devices nor special phase diffractive elements are required. In our set-up, we have recorded the digital hologram of two 2D objects located at different distances, and their images have been reconstructed evaluating the Fresnel diffraction integral in a computer. The reconstructions are shown in Fig. 1(d) and (e). These experimental results confirm the feasibility and flexibility of our method.

References

[1] Schnars U and Jüptner WPO 2002 *Meas. Sci. Technol.* **13** 85-101
[2] Martínez-León Ll, Araiza-E M, Javidi B, Andrés P, Climent V, Lancis J and Tajahuerce E 2009 *Opt. Express* **17** 12900-9

Acknowledgments

This work has been supported in part by the agreement between the *Universitat Jaume I* and the *Fundació Caixa Castelló (Bancaixa)*, grant P1-1B2006-29. Maria Araiza-E gratefully acknowledges a grant, 08I011.36/1, from the same agreement and partial suport by PROMEP PIFI 2007-33-07, Mexico. B. Javidi is grateful for support from Guggenheim Foundation. The authors are very grateful to the *Serveis Centrals d'Instrumentació Científica (SCIC)* of the *Universitat Jaume I* for the use of the lithography system based on direct laser writing.

Recent Progress and Perspectives in Digital Holographic Microscopy

P. Marquet[1,2] P. Jourdain[1], D. Boss[1,2], C. Depeursinge[3] and P. Magistretti[1,2]

[1]Centre de Neurosciences Psychiatriques, DP-CHUV, Prilly-Lausanne, Switzerland
[2]Ecole Polytechnique Fédérale de Lausanne, BMI, CH-1015 Lausanne, Switzerland
[3]Ecole Polytechnique Fédérale de Lausanne, Applied Optics Institute, Lausanne, Switzerland

Email : pierre.marquet@a3.epfl.ch

Abstract. Digital Holographic Microscopy (DHM), is a new imaging technique allowing to provide quantitative phase images with a high accuracy and stability making possible to explore a large variety of relevant processes, occurring on the µs to day time scale, in the fields including material research as well as cell biology. As a non invasive and real time imaging technique, DHM is particularly well suited for high throughput screening

1. Introduction

DHM we have developed [1,2, 3] is an imaging technique allowing both sub-wavelength axial resolution and real time observation capabilities. Pratically, an original numerical and robust numerical process [1], allows to reconstruct, from a single recorded hologram, in an off axis geometry, an exact replica of the wave diffracted by the observed object providing in particular an absolute phase measurement. In such a geometry, the use of a microscope objective, permits to form hologram with spatial frequencies adapted to the sampling capacity of a classical CCD camera and thus to obtain diffraction limited amplitude and phase reconstruted images (lateral resolution ≈ 360 nm with 1.4 NA objective). By using a standard camera, a single hologram acquisition can be achieved in a few ten of microseconds and down to two microseconds with fast cameras enabling acquisition rates of up to 10^5 frames/s. These extremely short acquisition times, associated with the absence of moving part, make DHM capable to monitor very fast processes. On the other hand, the numerical hologram reconstruction also allows to compensate for aberration [3], experimental noise (vibrations, thermal drift etc...) as well as to achieve an off-line numerical refocusing making it possible to measure, with a nanometric axial sensitivity and over several hours, quantitative phase mapping.

2. Applications

A large variety of applications, requiring reflection or transmission configuration, including material research as well as cell biology can be adressed with DHM. Stress evaluation by imaging birefringence, surface and interface sciences (roughness etc..), microtechnologies, micro-optics and MOEMS are important fields where DHM brings innovative and effcient solutions [4].
Concerning cell biology, the optical phase shift induced by a transparent living cell on the transmitted wavefront can be regarded as a powerful endogenous contrast agent, as it contains quantitative

information about both the morphology (cell volume, shape etc.) and the intracellular refractive index. Practically, DHM has demonstrated its capacity to efficiently and quantitatively investigating cell structure and dynamics [2,5]. Specifically, the DHM cell-induced phase shift has allowed the derivation of highly relevant cell parameters, including dry mass production, density and spatial distribution. Concretely, the evolution patterns of these parameters, have permitted to non-invasively characterize cell cycle at the single cell level [6]. On the other hand, technical development [7] have allowed to separately measure, from the phase signal, the information about cell morphology and about the intracellular refractive index, related to the intracellular content particularly the protein concentration and the water movement. As far as red blood cells (RBC) are concerned, such decomposition of the phase signal provides direct measurements of the mean corpuscular volume (MCV) and mean corpuscular hemoglobin concentration (MCHC) two parameters of great importance for diagnostic purposes. In addition, DHM high phase sensitivity has permitted to study the RBC membrane fluctuations, in particular their spatial distribution, providing information about the functional status of erythrocyte which can be linked to the body health state. Within the framework of neurobiology, a correlation between the intracellular refractive index variation and the transmembranar current has been established allowing a prediction of the actual cell current by a simple analysis of DHM phase shift.

3. Conclusion

DHM is a new quantitative imaging technique affording the possibility to address numerous issues in many application fields from material sciences to life sciences. Thanks to the phase information which can be related to very different biophysical cell parameters, including cell morphology and volume, intracellular content, cell membrane dynamics and mechanical properties, transmembranar current etc, DHM is a versatile and powerful non-invasive technique allowing to study very different aspects of cell dynamics, from the investigation of specific cell mechanisms to the development of solutions for high throughput screening.

References

[1] Cuche E, Marquet P et al. Appl Opt 1999; 38(34):6994-7001.
[2] Marquet P, Rappaz B et al. Opt Lett 2005; 30(5):468-70
[3] Colomb T, Montfort F et al. J Opt Soc Am A Opt Image Sci Vis 2006; 23(12):3177-3190.
[4] Montfort F, Emery Y et al., Proceeding of SPIE 2007; vol. 6463: 1-7
[5] Carl D, Kemper B et al. Appl Opt 2004; 43(36):6536-44.
[6] Rappaz B, Cano E et al. J. Biomed Opt 2009; 14(3):034049
[7] Rappaz B, Marquet P, et al. Opt Exp 2005;13(23):9361-9373
[8] Rappaz B, Barbul A, et al. Cytometry A 2008; 73(10): 895-903
[9] Rappaz B, Barbul A et al. Blood Cells Mol Dis 2009; 42(3) :228-32

Resolution in in–line Digital Holography

C Fournier[1,3], **L Denis**[1,2], **T Fournel**[1]

[1]Laboratoire Hubert Curien, CNRS UMR 5516, Université de Lyon, Université Jean-Monnet, 42000 Saint-Etienne.

[2]Ecole Supérieure de Chimie Physique Electronique de Lyon, 69616 Lyon, France.

[3]Telecom Saint Etienne, 42000 Saint Etienne.

E-mail: corinne.fournier@telecom-st-etienne.fr

Abstract. Digital in-line holography is a 3D imaging technique which has been widely developed during the last two decades. This technique achieves the 3D reconstruction of volume objects from a 2D image-hologram. It is a metrological tool and therefore the improvement of resolution is one of the current challenges. However the resolution depends on several experimental parameters and the experimenters have to choose the parameters which will lead to the best resolution. This paper presents the study of resolution in in–line digital holography from the asymptotical bounds of the covariance of estimators used in hologram reconstruction.

1. Introduction

Digital in-line holography is a metrological tool widely used in experimental mechanics, biology or fluid dynamics and therefore the improvement of resolution is one of the key issues of this field ([1], [2], [3], [4]). The resolution depends on several experimental parameters such as the sensor definition, recording distance, object size, location of the object in the field of view … and the experimenters have to choose the parameters which will lead to the best resolution. This paper describes an approach to the theoretical study of the resolution issue in in–line digital holography.

The concept of resolution is usually associated with the ability to distinguish two overlapping image-patterns of the same kind of objects. The resolution limit of lenses, microscopes, and telescopes are mostly based on the Rayleigh criterion (two-point resolution), which relates to the resolving power of the human eye to distinguish the images of two closely located point sources in observations of the sum of the images. Modern definitions of resolution, on the other hand, are based on statistical approaches. In this field two definitions of resolution have been distinguished [5]: (1) differential resolution (or two point resolution) is defined as the system's capability to determine the separation of two sources (Rayleigh's resolution), (2) the single-source resolution is defined as the system's capacity to determine the position of a point-source object that is observed in a background of noise. The latter is the most relevant in in-line digital holography, as the technique is mostly used to study volumes of isolated objects.

We use a statistical point of view to study single source resolution, and derive here the resolution on the optical axis.

2. Cramér-Rao lower bound on the covariance of maximum likelihood estimators

We will consider resolution from the view point of statistical parameter-estimation theory ([5], [6], [7]). The (optical) system response to a point source can be described by a parametrical model $g_{\boldsymbol{\theta}}(x_k, y_k)$, where $\boldsymbol{\theta}$ is a vector of parameters defining both the optical system and source location, and (x_k, y_k) are the pixel coordinates on the sensor. Due to noise, the measured image is a perturbed version of the model $g_{\boldsymbol{\theta}}(x_k, y_k)$. The estimation of the source location from a measured image can be performed based on a maximum likelihood (ML) criterion. Under a hypothesis of additive white Gaussian noise, this amounts to minimizing the squared residuals between measurements and model $g_{\boldsymbol{\theta}}$. ML estimators are unbiased and asymptotically efficient in the sense that their covariance reaches Cramér-Rao lower bound (CRLB). The covariance of any unbiased estimator is lower bounded by CRLB. The CRLB therefore provide an interesting insight of the achievable accuracy. Note that a ML approach has recently been introduced ([8], [9]) in particle digital holography.

3. Single-source resolution on the optical axis

We describe here the methodology of CRLB computation in the simple case of a point source located on the optical axis. The extension to a single source located outside of the optical axis is similar, but omitted here due to the paper length limit.

The diffraction pattern of a point source located at $\boldsymbol{\theta}(x, y, z)$ can be approximated by:

$$g_{x,y,z}(x_k, y_k) = \alpha.\sin\left(\pi \frac{(x - x_k)^2 + (y - y_k)^2}{\lambda z}\right) \tag{1}$$

Under the assumption of a white Gaussian noise, the analytical form of Fisher information matrix element $\left[\mathbf{I}(\boldsymbol{\theta})\right]_{i,j}$ is given [10] by:

$$\left[\mathbf{I}(\boldsymbol{\theta})\right]_{i,j} = \frac{1}{\sigma^2}\left(\frac{\partial \mathbf{g}}{\partial \theta_i}\right)^t \left(\frac{\partial \mathbf{g}}{\partial \theta_j}\right) \tag{2}$$

where $\mathbf{g}$ is a vector of components $g_{\boldsymbol{\theta}}(x_k, y_k)_{k=1..N}$. Ignoring the sampling on the sensor equation (2) can be rewritten as follows:

$$\left[\mathbf{I}(\boldsymbol{\theta})\right]_{i,j} = \frac{1}{\sigma^2}\frac{1}{L^2}\int_{-L/2}^{L/2}\int_{-L/2}^{L/2}\left(\frac{\partial g_{x,y,z}(x_k, y_k)}{\partial \theta_i}\frac{\partial g_{x,y,z}(x_k, y_k)}{\partial \theta_j}\right)\Bigg|_{\substack{x=0\\y=0}} dx_k dy_k \tag{3}$$

where L is the sensor width (considered square here). Using the fact that the point source is located on the optical axis $x = y = 0$ and the hypothesis L^2 □ λz, most often valid in in-line digital holography, Fisher matrix can be written:

$$\mathbf{I}(x, y, z) = \frac{\alpha^2}{\sigma^2}\begin{pmatrix} \dfrac{\pi^2 L^2}{6\lambda^2 z^2} & 0 & 0 \\ 0 & \dfrac{\pi^2 L^2}{6\lambda^2 z^2} & 0 \\ 0 & 0 & \dfrac{7\pi^2 L^4}{360\lambda^2 z^4} \end{pmatrix} \tag{4}$$

Fisher information matrix is diagonal and so is its inverse: the covariance matrix. Estimation errors on the parameters x, y and z are not correlated. The lateral and axial accuracy are given by the standard deviations:

$$\sigma_x = \sigma_y = c\frac{\sigma}{\alpha}\frac{\lambda}{L/z} \quad , \quad \sigma_z = 6\sqrt{5/21}c\frac{\sigma}{\alpha}\frac{\lambda}{(L/z)^2} \tag{5}$$

with $c = \sqrt{6}/\pi$.

This result is in agreement with classical formulas [11]: $\sigma_x \propto \frac{\lambda}{\Omega}$ and $\sigma_z \propto \frac{\lambda}{\Omega^2}$ and furthermore indicates that the resolution improves proportionally with the signal to noise ratio: $\frac{\alpha}{\sigma}$.

4. Conclusion :

This study describes a general methodology which can be used to derive the resolution or the estimation accuracy in digital holography. It is based on the computation of CRLB. This gives the classical relations for on-axis resolution. The resolution can also be computed outside of the optical axis. Let us note that the CRLB can be reached in practice using ML-based algorithms such as in ([8], [9]), provided that the global minimization problem can be solved. Sampling and quantization effects will also be modeled in a further work.

References

[1] M. Jacquot, P. Sandoz, et G. Tribillon, "High resolution digital holography," *Optics Communications*, vol. 190, 2001, pp. 87-94.

[2] A. Stern et B. Javidi, "Improved-resolution digital holography using the generalized sampling theorem for locally band-limited fields," *Journal of the Optical Society of America A*, vol. 23, 2006, pp. 1227–1235.

[3] J. Garcia-Sucerquia, W. Xu, S.K. Jericho, P. Klages, M.H. Jericho, et H.J. Kreuzer, "Digital in-line holographic microscopy," *Applied optics*, vol. 45, 2006, pp. 836–850.

[4] D.P. Kelly, B.M. Hennelly, N. Pandey, T.J. Naughton, W.T. Rhodes, et F. SPIE, "Resolution limits in practical digital holographic systems," *Optical Engineering*, vol. 48, 2009, p. 095801.

[5] A.J. Den Dekker et A. Van den Bos, "Resolution: a survey," *Journal of the Optical Society of America A*, vol. 14, 1997, pp. 547-557.

[6] M. Shahram et P. Milanfar, "Imaging below the diffraction limit: A statistical analysis," *IEEE Transactions on Image Processing*, vol. 13, 2004, pp. 677-689.

[7] C.W. Helstrom, "Resolvability of objects from the standpoint of statistical parameter estimation," *Journal of the Optical Society of America*, vol. 60, 1970, pp. 659–666.

[8] F. Soulez, L. Denis, C. Fournier, E. Thiébaut, et C. Goepfert, "Inverse problem approach for particle digital holography: accurate location based on local optimisation," *J. Opt. Soc. Am. A*, vol. 24, 2007.

[9] F. Soulez, L. Denis, É. Thiébaut, C. Fournier, et C. Goepfert, "Inverse problem approach in particle digital holography: out-of-field particle detection made possible," *Journal of the Optical Society of America A*, vol. 24, 2007, pp. 3708–3716.

[10] S.M. Kay, *Fundamentals of statistical signal processing: estimation theory*, 1993.

[11] J.W. Goodman, *Introduction to Fourier optics*, Roberts & Company Publishers, 2005.

Speckle Reduction Techniques in Digital Holography

David Monaghan[1], Damien Kelly[1], Bryan Hennelly[1], Bahram Javidi[2]

[1]Department of Computer Science, National University of Ireland, Maynooth, Co. Kildare, Ireland

[2]University of Connecticut Electrical & Computer Engineering Dept. 371 Fairfield Road, Unit 2157 Storrs, CT 06269-2157 USA

E-mail: bryanh@cs.nuim.ie

Abstract. We have studied several speckle reduction techniques, applicable to digital holography. These include the use of optical diffusers, wavelet filtering, simulating temporal incoherence and filtering in the Fourier domain. The Digital Holograms (DHs) used in this study are captured using a Phase Shift Interferometric (PSI) in-line setup and subsequently reconstructed numerically.

Digital holography [1], (DH) involves the use of discrete electronic devices, such as CCDs to record the hologram. Reconstruction may be performed either optically using a Spatial Light Modulator or numerically by simulating the propagation of the wave field back to the plane of the object. When coherent illumination is reflected from an optically rough surface, the resulting intensity pattern has a random spatial intensity variation known as speckle pattern. Speckle is a significant hindrance in imaging applications, such as DH, and acts to severely degrade the performance of coherent optical systems. Over the years many different techniques for reducing speckle have been developed, unfortunately however these techniques tend to reduce the resolution of the imaging system [3]. If several speckle fields are summed on an intensity basis the speckle contrast can be reduced significantly, see for example the discussion in Chap 5 [3]. Replaying holograms with a source that is not temporally coherent reduces speckle

Figure 1. (a) Original intensity, (b) after simulation of temporal incoherence, (c) after discrete Fourier filtering, (d) using wavelet thresholding

contrast. The process may be simulated by calculating the reconstruction at the correct distance

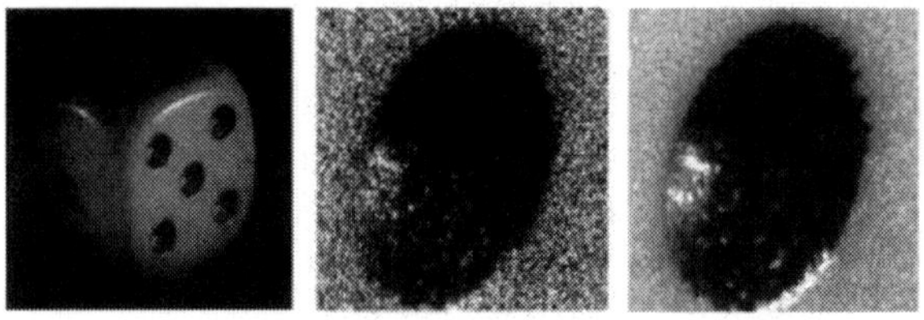

Figure 2. Speckle reduction using multiple diffuser illuminations.

for the hologram plane for a series of different wavelengths [4]. Each of these reconstructions is then summed on an intensity basis. We choose the difference in wavelengths such that the speckle patterns in the reconstructions are statistically independent. There is an blurring of the image itself when replaying at incorrect wavelengths which limits the number of statistically independent reconstructions that can be added together. The original reconstruction is shown in Fig. 1(a). Fig. 1 (b) shows the results 9 reconstructions have been added together on an intensity basis. The second technique we investigated is called Discrete Fourier Filtering [5], where the Fourier transform of the hologram is repeatedly spatially filtered and the resulting intensities are added together. Each spatial filter corresponds to a different rectangular band in the frequency domain. By moving the spatial filter in the Fourier domain we allow different parts of the hologram's Fourier spectrum to contribute to the holographic reconstruction. Again there is an inherent reduction in resolution. Fig. 1(c) shows the result 16 reconstructed segments have been added together on an intensity basis. In this case the final resolution is only one quarter that of the system. We also investigated wavelet filtering [6] on the reconstructed intensity. This method is the fastest and shows results that are at least as good as the other techniques but again there is a reduction in the resolution of the final image. Fig. 1(d) shows the results for an SWT wavelet transform, attained using a balance sparsity normal threshold applied by soft thresholding using a 'Haar' mother wavelet used with 3 levels. Our studies have shown that the only way to reduce speckle in digital hologram reconstructions without a reduction in resolution is to record multiple holograms with different diffuse illumination and then sum the intensities of the reconstructions. In Fig. 2(a) we show the result of summing 35 such intensities together. A zoomed in comparison of before and after this improvement is shown in In Fig. 2(b) and (c)

Acknowledgments
This research is funded from the European Community's Seventh Framework Programme FP7/2007 2013 under grant agreement 216105 "Real 3D".

References

[1] Goodman J.W., Lawrence R. 1967 Digital image formation from electronically detected holograms *Appl. Phys. Lett.* **11** pp 77-79

[2] Hennelly B.M., Sheridan J.T. 2005 Generalizing, optimizing, and inventing numerical algorithms for the fractional Fourier, Fresnel, and linear canonical transforms, *J. Opt. Soc. Am. A* **22** pp 917-927

[3] Goodman J.W. 2007, *Speckle Phenomena in Optics*, Roberts Company (Colorado)

[4] Hennelly B.M., Kelly D.P., Maycock J., Naughton T.J., McDonald J.B. 2007 Speckle Reduction from Digital Holograms by Simulating Temporal Incoherence *SPIE Proc.* **6696**

[5] Maycock J., Hennelly B.M., McDonald J.B., Frauel Y, Castro A., Javidi B., Naughton T.J., Reduction of speckle in digital holography by discrete Fourier filtering *J. Opt. Soc. Am. A* **24** pp 1617-1622

[6] Molony K.M, Maycock J., Mc Donald J.B., Hennelly B.M., Naughton T.J. 2008 A comparison of wavelet analysis techniques in digital Holograms," *SPIE Proc.* **6694**

Journal of Physics: Conference Series **206** (2010) 012027 doi:10.1088/1742-6596/206/1/012027

Zooming Algorithms for Digital Holography

Bryan Hennelly, Damien Kelly, Nitesh Pandey, David Monaghan
Department of Computer Science, National University of Ireland, Maynooth, Co. Kildare, Ireland
E-mail: bryanh@cs.nuim.ie

Abstract. Digital Holography is an imaging modality made up of two parts: (i) Recording an interference pattern on a CCD, where an object wave field and a known reference wave are coincident and extracting the complex object wave field from this interference pattern and (ii) Replaying or reconstructing the hologram on a computer by simulating the propagation of the object wave field back to the object plane. Thus an image is obtained. We show how to adapt the reconstruction algorithm in a simple way to allow it to generate any output range and in any location making it far more versatile for zooming in on specific regions of our reconstructed image.

Digital holography [1], (DH) involves the use of discrete electronic devices, such as CCDs to record the hologram. In this case reconstruction is performed numerically by simulating the propagation of the wave field back to the plane of the object [2]. The most commonly used algorithms for the reconstruction algorithm are the direct method and the spectral method of calculating the Fresnel Transform which describes free space propagation in the paraxial approximation. These algorithms differ in the output range that they display. For the direct method the output image size is proportional to the propagation distance making it more appropriate for large objects at large distances. The spectral method has an output image size equal to the size of the CCD. In this study we have developed an algorithm that allows us to arbitrarily change the output pixel size and therefore the output window size as we see fit, we therefore can zoom in on particular features or areas of our reconstructed image.. The basic principle is based on the following property of the Fresnel transform, which is denoted by the operator F_{z_n}.

$$u_z(x/M) = \exp\left(j\frac{2\pi}{\lambda f_2}x'^2\right)F_{z_n}\{u(x')\exp\left(j\frac{2\pi}{\lambda f_1}x'^2\right)\} \quad (1)$$

where M is a magnification parameter that we are free to choose and this imposes the conditions;

$$z_n = zM, f_1 = z_n/\left(M-1\right), f_2 = z_n - f_1 \quad (2)$$

Equation 1 implies that to obtain a magnified image of our reconstructed image, our digital hologram can be multiplied by a discrete chirp signal with parameter f_1 and then reconstructed using the spectral method where we propagate a distance z_n and finally we multiply by another discrete chirp function. The user can define the output sampling interval arbitrarily, which will in turn define values for M, z_n, f_1 and f_2. In the case of the direct method we cannot apply this idea successfully to zoom because the terms cancel to reveal an output width that is a constant. We note that the relationship given in Equation 1 was first used to create a zooming algorithm for the Fresnel transform, known as the 'Rhodes Light Tube' in [3] where the output sampling interval was set to be a constant for all distances. When applying a large magnification to

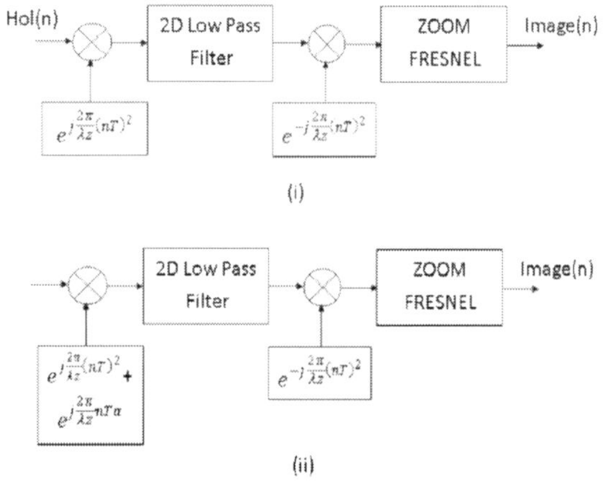

Figure 1. Flowcharts for adapting the zoom algorithm for magnifying the object: (i) Using a low pass filter between two chirps will mean that energy from the signal outside the magnified region that we wish to image will be set to zero and will no longer be a source of noise; (ii) we can introduce a linear phase to make the low pass filter into a bandpass filter and thus we can magnify different regions of the object

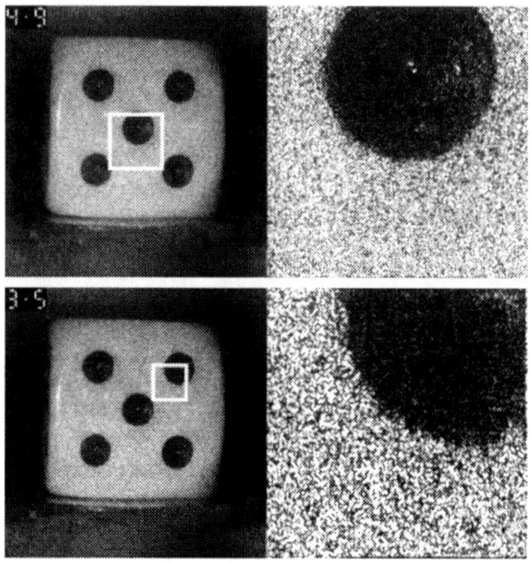

Figure 2. Zooming in with the adapted low pass filtering technique implemented before applying the zoom algorithm. In this case we also show the effect of using a linear phase at the input to scan the object. The numbers on the figure indicate the width of the image in mm

the algorithm such that we can view small parts of the object we find that the algorithm fails because those parts of the object that we are not interested in wrap around and invade the space of our field of interest. In order to overcome this we have developed a procedure shown in Fig. 1 and 2 where by applying an appropriate low pass filter we can eliminate this effect allowing to zoom into an object and view with any pixel size we wish. The actual resolution that we can still detect will be determined by the point spread function of the system. We believe this algorithm will have important application in digital holographic micrpscopy.

0.1. Acknowledgments
This research is funded from the European Community's Seventh Framework Programme FP7/2007 2013 under grant agreement 216105 "Real 3D".

References
[1] Goodman J.W., Lawrence R. 1967 Digital image formation from electronically detected holograms *Appl. Phys. Lett.* **11** pp 77-79
[2] B. M. Hennelly, J. T. Sheridan 2005 Generalizing, optimizing, and inventing numerical algorithms for the fractional Fourier, Fresnel, and linear canonical transforms, *J. Opt. Soc. Am. A* **22** pp 917-927
[3] W. T. Rhodes 2001 Numerical simulation of Fresnel-regime wave propagation: the light tube model, in *Wave-Optical Systems Engineering, F. Wyrowski, ed., Proc. SPIE* **4436** pp 2126
[4] X. Deng, B. Bihari, J. Gan, F. Zhao, R. T. Chen 2000 Fast algorithm for chirp transforms with zooming-in ability and its applications *J. Opt. Soc. Am. A* **17** pp 762-771
[5] F. Zhang, I. Yamaguchi, L. P. Yaroslavsky 2004 Algorithm for reconstruction of digital holograms with adjustable magnification *Opt. Lett.* **29** pp 1668-1670

Deformation of digital holograms for full control of focus and for extending the depth of field

M. Paturzo, P. Memmolo, A. Finizio and P.Ferraro

CNR–Istituto Nazionale di Ottica Applicata and Istituto di Cibernetica, via Campi Flegrei 34, 80078 Pozzuoli (NA), Italy

E-mail: pietro.ferraro@inoa.it

Abstract. We present a new method to manage the depth of focus in holograms numerical reconstructions through an adaptive deformation of the digital holograms. We demonstrate that this technique can be applied both to Fourier and Fresnel holograms. The experimental results, in agreement with the theoretical model, are shown and commented for both these configurations.

1. Fresnel holograms

Considering a digital hologram of an object at distance d from the CCD, the in focus numerical reconstruction is obtained adopting the numerical modelling of the diffraction Fresnel propagation integral. If an affine geometric transformation, a deformation, is applied to the original recorded hologram h, consisting in a simple stretching described by $[\xi' \eta'] = [\xi \quad \eta \quad 1]T$ through the operator

$$T = \begin{bmatrix} \alpha & 0 \\ 0 & \alpha \\ 0 & 0 \end{bmatrix} \tag{1}$$

given by $h(\xi', \eta') = h(\alpha\xi, \alpha\eta)$ we obtain that the propagation integral changes as follows,

$$B(x,y,d) = \frac{1}{i\lambda d} e^{ikd} \iint h(\alpha\xi, \alpha\eta) e^{\frac{ik\alpha^2(x-\xi)^2}{2d\alpha^2}} e^{\frac{ik\alpha^2(y-\eta)^2}{2d\alpha^2}} d\xi d\eta =$$

$$= \frac{1}{i\alpha^2 \lambda d} e^{ikd} \iint h(\xi', \eta') e^{\frac{ik(x'-\xi')^2}{2D}} e^{\frac{ik(y'-\eta')^2}{2D}} d\xi' d\eta' = \frac{1}{\alpha^2} b(x', y', D) \tag{2}$$

From Eq.(2) it is clear that the new hologram reconstructs the object in focus at a different distance $D = \alpha^2 d$. The hologram has been simply stretched and we name α the elongation factor. The impact that the stretching has on numerical reconstructions, is shown in Fig.1. The object was made of three different wires positioned at different distances from the CCD array. The horizontal wire, the twisted wire with the eyelet and the vertical wire were set at distances of 100mm, 125mm and 150mm, respectively. The numerical reconstructions at the above distances give an image in which each wire at a time is in good focus, as shown in Fig.1a, 1b and 1c, at the corresponding recording distances. When the hologram h(x,y) is uniformly stretched with an elongation factor along both dimensions (x,y) of

α=1.1, the horizontal wire results to be in focus at a distance d =150 mm (Fig.1d) instead of 100mm. If α=1.22, the twisted wire with the eyelet is in focus at d= 150 mm (Fig.1e)instead that at d=125mm. Anyway, the results shown in Fig.1 demonstrates that the depth of focus can be controlled by means of the uniform stretching of the holograms according to Eq.(2).

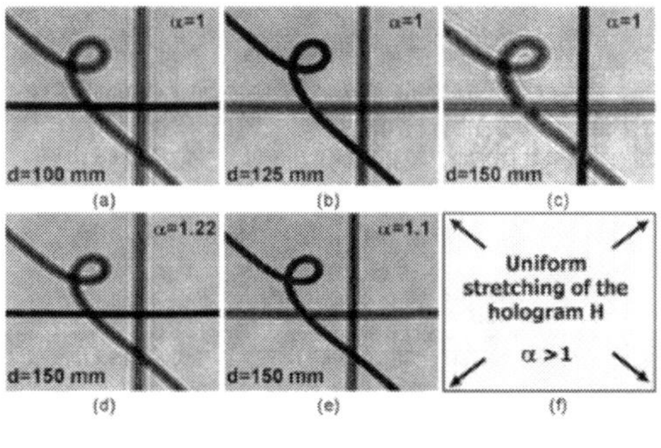

Figure 1. Reconstruction of a digital hologram H as recorded (no-stretching) at three different distances with each wire in focus at distance of (a) 100mm, (b) 125mm and (c) 150 mm, respectively. If the holograms is stretched with α=1.1 the horizontal wire is in focus at distance of d =150 mm (d) while for α=1.22 the curved wire is in focus even at d= 150 mm (e).Conceptual draw of the H stretching (f).

This interesting results implies that by an adaptive deformation of digital holograms, different parts of the 3D scene that are a different depths can be obtained in good focus in a single reconstructions.

2. Fourier holograms

For holograms recorded in Fourier configuration, the simple linear stretching gives different results in respect to the Fresnel holograms. In fact, by application of a linear transformation to a Fourier hologram no changes can be observed about image focusing properties during reconstruction.

Nevertheless, if, instead of a linear transformation described by Eq.(1), we apply a quadratic deformation, it results that the in-focus distance changes with a linear law given by $D = L(1 + 2\alpha l')$ where l' is the horizontal coordinate in the reconstruction plane, α is the deformation parameter and L is the reconstruction distance of the recorded (not deformed) hologram. Therefore, by this kind of hologram deformations we can put correctly in-focus the numerical reconstruction of a tilted object recovering its EFI. In Fig.2 (left) we show the numerical reconstruction of the acquired hologram for an object tilted with an angle φ= 55°. It is clear that the left part of the object, where there is the number "0", results to be in-focus, while the right part, with the number "1", is out of focus and that the focusing gradually worsen going from left to right side. If we apply the quadratic deformation before performing the reconstruction (with an opportune choice of the deformation parameter α), we obtain the image shown in Fig.2(right).

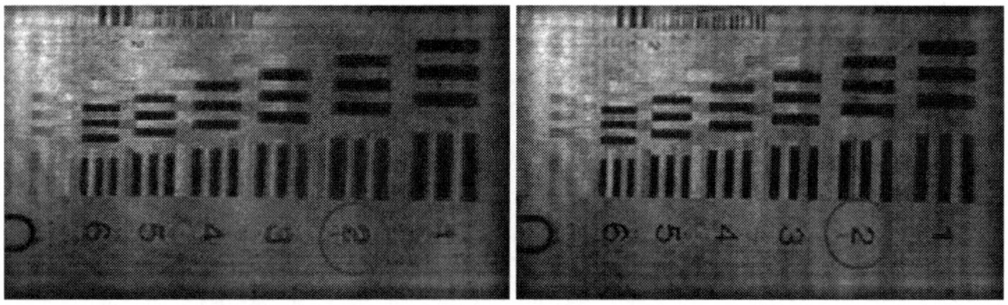

Figure 2. Numerical reconstruction of the holograms for an object tilted with an angle of 55 degrees as acquired by the CCD (left) and after the quadratic deformation (right).

Reference

[1] Ferraro P, Paturzo M, Memmolo P and Finizio 2009 A *Opt. Lett.* **34**, 2787
[2] Paturzo M and Ferraro P 2009 *Opt. Express* **17** 20546

Journal of Physics: Conference Series **206** (2010) 012029 doi:10.1088/1742-6596/206/1/012029

Improvement of Viewing-Zone Angle and Image Quality of Digital Holograms

Takanori Nomura

Faculty of Systems Enigneering, Wakayama Univesity, 930 Sakaedani, Wakayama, 640-8510, Japan

E-mail: `nom@sys.wakayama-u.ac.jp`

Abstract. The method to improve of a viewing-zone angle and an image quality of a digital hologram is presented. A number of digital holograms of a central object are recorded from the position on the circumference. The holograms are used for a hologram synthesis to improve the image quality from whole viewing-zone angle. The synthesis is achieved by a correlation between a hologram and numerically propagated holograms. The large-sized synthesized digital hologram has a wide viewing-zone angle and less speckles. Some experimental results are shown to confirm the proposed method.

1. Introduction

Improvement of the quality of the reconstructed image in digital holography is necessary, because the image is suffering from speckles. Some approaches [1, 2, 3] have been proposed to improve the quality in terms of reduced speckles. However, the viewing-zone angle defined by the size of the hologram and the distance from object to hologram is small, because the area of imaging device is much smaller than a conventional film hologram. Furthermore, it is useful for 3D observation if a digital hologram from any viewing-zone angle is obtained. To achieve the wide viewing-zone angle, a mathematical approach of superposition of holograms has been proposed [4]. In the method, some non-overlapped digital holograms are used for superposition. However for perfect superposition, the difficulty in knowing exactly the amount of rotation remains. In this study, the straightforward method to superpose digital holograms recorded from various viewing-zone angles on cncyclic positions is proposed.

2. Recording and synthesis of digital holograms

Digital holograms of the object, which is placed on the rotational stage, from different angles are recorded by an image sensor concyclic positions as shown in Fig. 1. If N holograms are recorded at angular intervals of $2\pi/N$ radian, a reconstructed image is obtained at angular intervals of $2\pi/N$ radian. Therefore, wide viewing-zone angle obtained. However, the quality of the reconstructed image is very poor, because it is suffering from the speckles due to the small extent of the hologram. To reduce the speckles, it is effective to use a hologram with a large extent [1]. For example, if M holograms are recorded to be synthesized into a large extent hologram for each N digital hologram, a good quality image will be reconstructed. In this case, MN holograms should be recorded. This is not practical. In the recording process, the concerned hologram and neighboring holograms are obtained. The neighbors are not on the

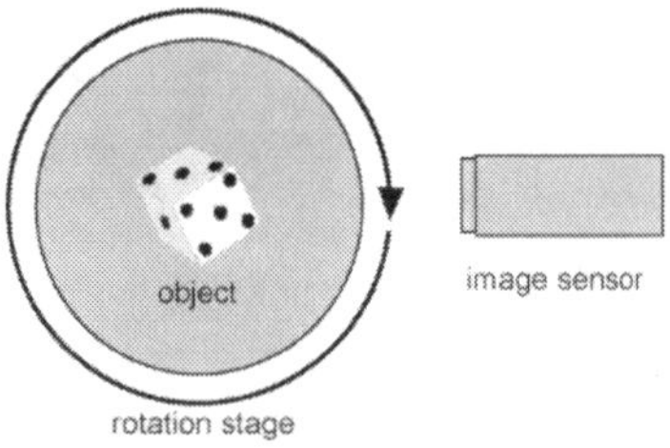

Figure 1. Recording digital holograms by an image sensor on the concyclic positions.

same plane of the concerned hologram. To obtain the hologram on the same plane where the concerned hologram lays, a technique is introduced. It is assumed that only phase is changed according to the propagation distance. Using this assumption, the inclination of the holograms are compensated.

3. Experimental results

Experimental results are given to confirm the feasibility of the proposed method. The object was rotated from −180 deg to +180 deg at an interval of 0.2 deg, and 1801 holograms were recorded. Then, some of recorded holograms were processed and superposed by the proposed method. The reconstructed images are shown in Fig. 2. Figure 3 shows the reconstructed

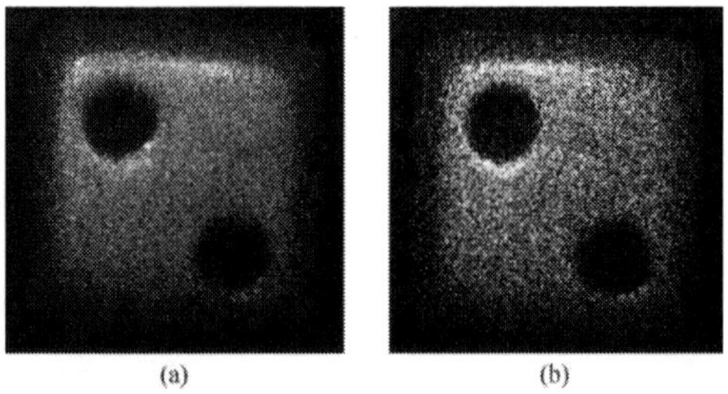

Figure 2. Reconstructed images from (a) a synthesized hologram and (b) a single hologram.

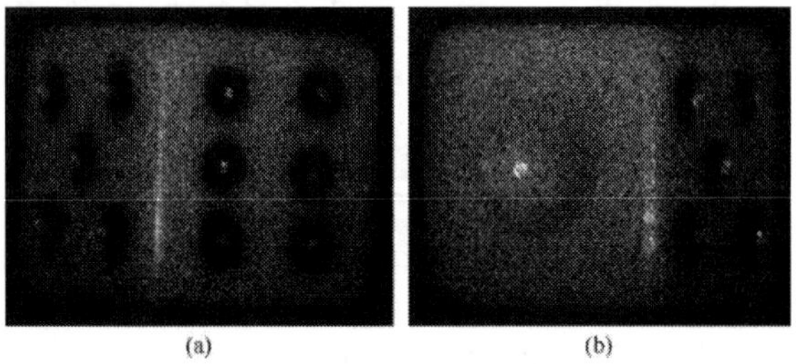

Figure 3. Reconstructed images from different viewing angles. (a) from left side (60 deg) and (b) from right side (60 deg).

images from different viewing directions. Different aspects of the object can be seen in the figures. The reconstruction image pair can be used for binocular stereopsis.

Acknowledgements

This study was partly supported by a research grant form The Murata Science Foundation.

References

[1] Binet R, Colineau J, and Lehureau J. -C 2002 *Appl. Opt.* **41** 4775–4782
[2] Nomura T, Okamura M, Nitanai E, and Numata T 2008 *Appl. Opt.* **47,** D38–D43
[3] Nakatsuji T and Matsushima K 2008*Appl. Opt.* **47,** D136–D143
[4] Hennelly B. M, Mc Elhinney C, Frauel Y, Naughton T. J, McDonald J. B, and Javidi B 2006 *AIP Conf. Proc.* **860,** 281–289

Stereo perception of reconstructions of digital holograms of real-world objects

Taina M. Lehtimäki[1], Kirsti Sääskilahti[1], Risto Näsänen[1] and Thomas J. Naughton[2,1]

[1] University of Oulu, Oulu Southern Institute, Ylivieska, Finland

[2] Department of Computer Science, National University of Ireland Maynooth, Ireland

E-mail: taina.lehtimaki@oulu.fi

Abstract. In digital holography a 3D scene is captured optically and often the perspectives are reconstructed numerically. In this study we digitally process the holograms to allow them to be displayed on autostereoscopic displays. This study is conducted by subjective visual perception experiments comparing single reconstructed images from left and right perspective to the resulting stereo image.

1. Introduction

In contrast to conventional holography, in digital holography [1] we often capture optically a 3D scene and reconstruct the desired perspectives numerically in a computer. The reconstructions are routinely displayed in the form of a 2D image slice through the reconstruction volume, an extended focus image, or a depth map from a single perspective. These are fundamentally 2D (or at most 2.5D) representations and for some scenes are not certain to give the human viewer a clear perception of the 3D features encoded in the hologram (occlusions are not overcome, for example). In addition digital numerical reconstructions are limited by speckle noise and limited depth of field.

Alternatively, an optical reconstruction would allow the human viewer to extract the 3D information themselves through stereopsis and motion parallax, but convincing full-field digital holographic displays for real-world 3D scenes have not been demonstrated due to current limitations of programmable display devices.

As an intermediate measure, we have digitally processed the holograms to display them on conventional autostereoscopic displays, which provide a readily understood and off-the-shelf available technology, and the human visual system itself extracts the 3D information of the scene encoded in the hologram. Previously we have reported results suggesting that when digital holographic reconstructions are displayed stereoscopically people perceive less speckle noise and increased depth of field [2,3]. In this paper we show more detailed results how displaying the reconstructions in stereo reduces the perception of noise, increases the detail visibility and general enhances perceived quality when compared to displaying the reconstructions as 2D images. Also we will show how noise perception changes as more hologram data (larger window) contributes to the reconstruction.

2. Method – subjects, apparatus, stimuli, procedure

The experiment was performed with thirteen subjects with normal or corrected-to-normal vision. Stimuli were presented on autostereoscopic display (Sharp LL-151-3D, 1024×768 pixels in total). The viewing distance was approximately 60 cm. We presented subjects with reconstructions from five different holograms, using six different window sizes (apertures on the hologram data) and two perspectives (left and right) from each hologram. We asked subjects to evaluate reconstructions of individual left and right perspectives, and resulting stereo pairs. Subjects answered either yes or no to questions: "Does noise disturb?" and "Are details visible?" They also evaluated the quality of images on scale 1–5 (1= very poor, 5=very good quality).

3. Results and discussion

The results from the experiment are shown in Figs. 1(a) and 1(b) as percentage of "yes" responses, and in Fig. 1(c) as an average over the thirteen subjects and five holograms. Compared to single images, Figure 1(a) shows that there is less disturbing noise in stereo images, Fig. 1(b) shows that more details are visible in stereo, and Fig. 1(c) shows that stereo images are perceived as being of higher quality.

The limitation on these results is that they are not quantified by psychophysical methods. Perceptual performance can be effectively measured by psychophysical methods such as threshold estimation techniques. Psychophysical threshold estimation methods produce highly accurate and quantitative ratio scale level descriptions of human performance, for example, the minimum amount of disparity that can be discerned with a given probability or the minimum time required for the recognition of a 3D object. In order to further investigate enhanced quality and noise reduction phenomena described here and extended focus phenomena described previously [2,3] we need to measure both perceived quality and perceived "stereo effect" by psychophysical methods. We propose to quantify binocular depth fusion by doing a specific experiment with increased depth of field stimuli.

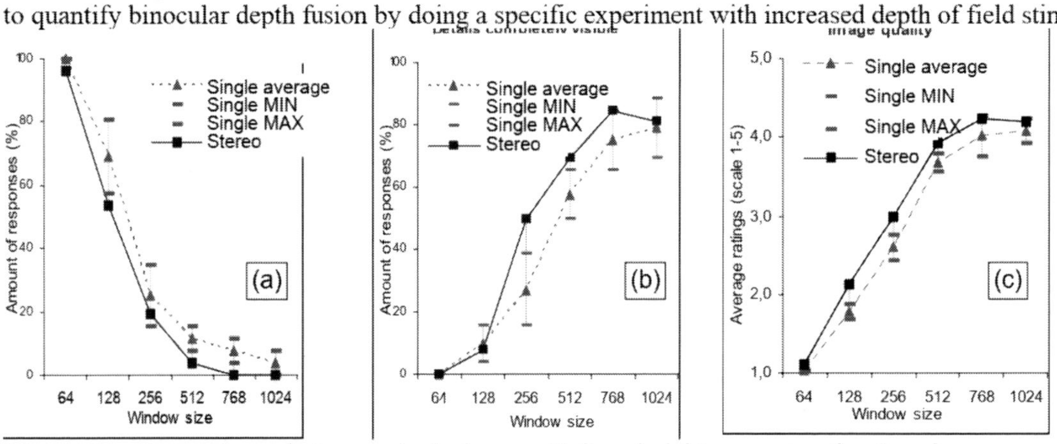

Figure 1. Comparisons between single images (left and right reconstructions) and corresponding stereo images for different window sizes. MIN corresponds to single left or right image that got the least "yes" responses (MAX is most responses). On vertical axis (a) percentage of "yes" responses to: "Does noise disturb substantially?" (b) percentage of "yes" responses to: "Are details completely visible?", and (c) Evaluated quality using scale 1-5 (1= very poor, 5=very good quality).

Acknowledgements
We thank Conor Mc Elhinney for capturing the digital holograms. We acknowledge support from Academy of Finland, Science Foundation Ireland under the National Development Plan, a European Commission Marie Curie Intra-European Individual Fellowship, and the European Community's Seventh Framework Programme FP7/2007–2013 under grant agreement no. 216105 (Real 3D).

References

[1] Th. Kreis, Handbook of Holographic Interferometry: Optical and Digital Methods, Wiley, 2005.
[2] T.M. Lehtimäki and T.J. Naughton, "Stereoscopic viewing of digital holograms of real-world objects," 3DTV-Conference 2007, IEEE Press, pp. 1–4, Kos, Greece, 7–9 May 2007.
[3] T.M. Lehtimäki, K. Sääskilahti, R. Näsänen, and T.J. Naughton, "Visual perception of digital holograms on autostereoscopic displays," Proc. SPIE vol. 7329, article no. 73290C, 2009.

Multispectral image pansharpening based on the contourlet transform

Israa Amro and Javier Mateos

Departamento de Ciencias de la Computación e I.A. , Universidad de Granada, 18071 Granada, Spain

E-mail: iamro@correo.ugr.es, jmd@decsai.ugr.es

Abstract. Pansharpening is a technique that fuses the information of a low resolution multispectral image (MS) and a high resolution panchromatic image (PAN), usually remote sensing images, to provide a high resolution multispectral image. In the literature, this task has been addressed from different points of view being one of the most popular the wavelets based algorithms. Recently, the contourlet transform has been proposed. This transform combines the advantages of the wavelets transform with a more efficient directional information representation. In this paper we propose a new pansharpening method based on contourlets, compare with its wavelet counterpart and assess its performance numerically and visually.

1. Wavelet and Contourlet-based pansharpening

In the literature many pansharpening methods have been proposed for combining PAN with MS image, see [1, 2] for a detailed review. It was found that multi-sensor image fusion is a trade-off between the spectral information from an MS sensor and the spatial information from an PAN sensor that is easily controlated with the wavelet transform fusion methods. However, wavelets fail to capture the smoothness along the contours [3]. The contourlet transform, an alternative multiresolution approach, provides an efficient directional representation and also efficient in capturing intrinsic geometrical structures of the natural image along the smooth contours [4]. A wide amount of methods had been proposed both for wavelets and contourlets. See [5, 6] for a detailled comparison of wavelet based fusion methods. Contourlets methods based on those wavelets methods has been also proposed in the literature. Also, the WiSpeR method can be understood as a generalization of different wavelet-based image fusion methods [7]. It uses a modification of the non-subsampled additive wavelet algorithm where the contribution from the PAN image to each of fused bands depends on a factor generated both from the sensor spectral response and physical properties of the observed object. We proposed a new contourlet panshapening method, named CiSpeR, that, similarly to WiSpeR, depends of a spectral factor to determine the amount of spatial detail of the PAN image that has to be injected into each MS band but it uses the non-subsampling contourlet transform with necessary filters, and n resolution levels and m directions in each level.

2. Experimental Results

The contourlet-based and wavelet-based methods mentioned in section 1 are evaluated on the dataset acquired by SPOT5 satellite shown in Fig. 1. The pansharpened images resulted from

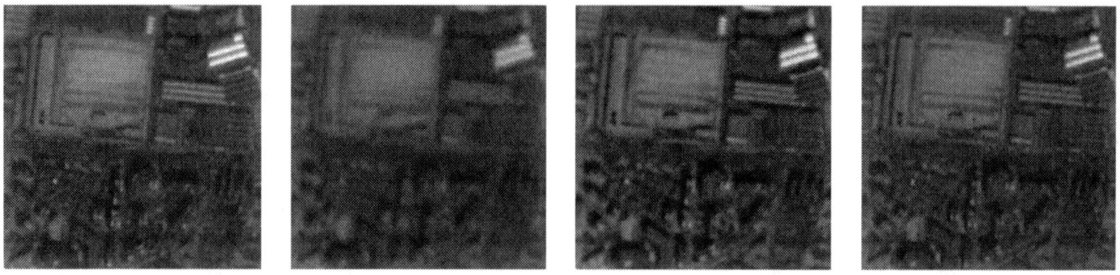

Figure 1. From left to right: SPOT-5 PAN image, low resolution image formed from the G-R-NIR bands of the MS SPOT image and pansharpened images using WiSPeR and CiSPer.

	COR (Ideal Value: 1)				UIQI (Ideal Value: 1)				ERGAS
	band 1	band 2	band 3	band 4	band 1	band 2	band 3	band 4	
WiSpeR	0.90	0.92	0.93	0.88	0.79	0.89	0.85	0.85	7.29
CiSpeR	**0.92**	**0.94**	**0.95**	**0.92**	**0.91**	**0.97**	**0.95**	**0.94**	**2.86**

Table 1. SPOT 5 Quantative Analysis

using the wavelet and contourlet methods under study are presented in Fig. 1. From the pansharpened images we observe that the proposed method, CiSpeR, enhances the images spatially while accurately preserves the spectral information. Numerical results in Table 1 agree with visual inspection. The highlighted values in the table present the highest (close to ideal ideal) values for each measure. It is clear that the contourlet-based method has better results than wavelet one. This result is expected since the contourlet is known to have a better representation for directional information.

3. Conclusions

In this paper, wavelet and contourlet based pansharpening approaches have been compared and their efficiency to merge SPOT images has been evaluated visual and quantitatively. The contourlet-based pansharpened images present, visually and numerically, better results than those obtained by wavelet, extracting spatial information from the PAN image missing in the MS image, without modifying its spectral information content.

Acknowledgments

This work has been supported by the Consejería de Innovación, Ciencia y Empresa of the Junta de Andalucía under contract P07-TIC-02698.

References

[1] Pohl C and Genderen J L V 1998 *Int. J. Remote Sens.* **19** 823–854
[2] Alparone L, Wald L, Chanussot J, Gamba P and Bruce L M 2007 *IEEETrans. Geosci. Remote Sens.* **45** 3012–3021
[3] Shah V P, Younan N H and King R 2007 *Proc. IEEE Int. Geosci. Remote Sens. Symp. IGARSS 2007* pp 310–313
[4] Do M N and Vetterli M 2005 *IEEE Trans. Image Process.* **14** 2091–2106
[5] González-Audícana M and Otazu X 2005 *Int. J. Remote Sens.* **26** 595–614
[6] Nuñez J, Otazu X, Fors O, Prades A, Pala V and Arbiol R 1999 *IEEE Trans. Geosci. Remote Sens.* **37** 1204–1211
[7] Otazu X, Gonzlez-Audcana M, Fors O and Núñez J 2005 *IEEE Trans. Geosci. Remote Sens.* **43** 2376–2385

2009 Euro-American Workshop on Information Optics IOP Publishing
Journal of Physics: Conference Series **206** (2010) 012032 doi:10.1088/1742-6596/206/1/012032

Mathematical Morphology and Modeling of Random Media

Dominique Jeulin

Centre de Morphologie Mathématique, Mathématiques et Systèmes, Mines ParisTech, 35, rue St-Honoré, F77300 Fontainebleau, France

E-mail: `dominique.jeulin@mines-paristech.fr`

Abstract. This paper is a short introduction to the modeling of complex microstructures by models of random sets, and to their use for predicting the effective properties of materials by means of numerical simulations.

1. Introduction

Since the seminal work of G. Matheron on porous media [11], and thanks to its implementation by means of image analysis [15], many applications based on mathematical morphology were developed. In this lecture, we introduced some models of random sets, and their use in the solution of homogenization problems.

2. Models of random sets

An efficient way to model random media is based on the theory of random sets [11, 13], giving a strong theoretical framework for modeling heterogeneous microstructures by means of a probabilistic approach. A random closed set A is characterized by means of its Choquet capacity $T(K)$, defined on the compact sets K, as introduced by G. Matheron [11, 13]:

$$T(K) = P\{K \cap A \neq \varnothing\} = 1 - P\{K \subset A^c\}$$

The probability depending on A and on K can be theroretically calculated on some models, and can be estimated on real or on simulated media by image analysis [3, 4, 11, 15, 16]. This gives the link between theory and applications, and the way to estimate the parameters of a model from images.

Models of random sets provide representations of microstructures and ways for simulations, in order to predict morphological as well as physical properties of complex media. This purpose is illustrated by some basic models (using a combination of a Poisson point process and of random grains) and their applications to real data: the Boolean model [11, 13, 15, 16, 4], the dead leaves model [3, 4], and multiscale versions obtained from a Cox point process [6, 7, 8]. They can show very different percolation thresholds [6, 7], which control their transport or physical properties. Other models, derived from solutions of reaction-diffusion equations, provide striking natural textures [2].

3. Homogenization

The theory of random sets also provides tools to solve homogenization problems, in order to predict the macroscopic properties of random media [11, 1, 10, 5]. Variational techniques give bounds of effective properties of random media, depending on a limited amount of information on the microstructure. The availability of 3D images, e.g. obtained by microtomography or generated by simulations of realistic models of microstructures, combined to solvers of PDE, gives access to the direct estimation of the effective properties, instead of bounds. The elastic moduli and the thermal conductivity can be obtained by finite elements [9]; using iterations of FFT [14], it is possible to solve in 3D the equations of electrostatics or of elasticity on images without meshing the microstructure; this was applied to the estimation of the dielectric permittivity of composites [8] and of the elastic bulk modulus of a porous Boolean models containing spherical pores or rigid inclusions, and therefore showing an infinite contrast [17]. For all these numerical estimations on finite domains must be solved the question of the RVE (Representative Volume Element), which we address by a statistical approach [9, 17]. Based on the use of the integral range introduced in Geostatistics [12], it provides intervals of confidence of the estimates of properties depending on the microstructure, the properties and the contrast of components. Numerical homogenization of random sets opens the way to the optimization of microstructures with respect to given physical properties.

References

[1] Beran M J 1968 *Statistical Continuum Theories* (New York: J. Wiley)
[2] Decker L and Jeulin D 1999 3D Spatial time structure simulations by reaction-diffusion models *Acta Stereologica* **18** (2) 247-54
[3] Jeulin D 1997 Dead Leaves Models: from space tesselation to random functions *Proc. of the Symposium on the Advances in the Theory and Applications of Random Sets (Fontainebleau, 9-11 October 1996)* ed D Jeulin (Singapore: World Scientific) pp 137-156
[4] Jeulin D 2000 Random texture models for materials structures *Statistics and Computing* **10** 121-31
[5] Jeulin D 2001 Random Structure Models for Homogenization and Fracture Statistics *Mechanics of Random and Multiscale Microstructures (CISM Lecture Notes N° 430)* ed D Jeulin and M Ostoja-Starzewski (Wien: Springer Verlag) pp 33-91
[6] Jeulin D and Moreaud M 2005 Multi-scale simulation of random spheres aggregates-Application to nanocomposites *Proc. 9th European Congress for Stereology (Zakopane, Poland), May 10-13, 2005)* ed J Chraponski, J Cwajna, L Wojnar
[7] Jeulin D and Moreaud M 2006 Percolation of multi-scale fiber aggregates *Proc. S4G, 6th Int. Conf. Stereology, Spatial Statistics and Stochastic Geometry (Prague, 26-29 June 2006)* ed R Lechnerova, .Saxl, V.Benes (Union Czech Mathematicians and Physicists) pp 269-74.
[8] Jeulin D and Moreaud M 2006 From Nano to Macro: a multiscale morphological approach of the dielectric permittivity of carbon-black nanocomposites *Invited Keynote Lecture to 10th Japanese-European Symposium on Composite Materials, Interactive Solution on Nano/Macroscopic Level and Hybrid Technology for Practical Applications (Nagano, Japan, Sept 26-28, 2006)* pp 1-5
[9] Kanit T, Forest S, Galliet I, Mounoury V, and Jeulin D 2003 Determination of the size of the representative volume element for random composites: statistical and numerical approach *Int. J.Solids Struct.* **40** (13-14) 3647–79
[10] Kröner E 1971 *Statistical Continuum Mechanics* (Berlin: Springer Verlag)
[11] Matheron G 1967 *Eléments pour une théorie des milieux poreux* (Paris: Masson)
[12] Matheron, G 1971 *The theory of regionalized variables and its applications* (Paris School of Mines)
[13] Matheron G 1975 *Random sets and integral geometry* (New-York: J. Wiley)
[14] Moulinec H and Suquet P 1994 A fast numerical method for computing the linear and nonlinear mechanical properties of composites *C.R. Acad. Sci. Paris* **318**, Série II 1417-23
[15] Serra J 1982 *Image analysis and mathematical morphology* (London: Academic Press)
[16] Stoyan D, Kendall WS and Mecke J 1987 *Stochastic Geometry and its Applications* (New-York: J. Wiley)
[17] Willot F and Jeulin D 2009 Elastic behavior of materials containing boolean random sets of inhomogeneities *Int. J. Eng. Sci.* **47** 313–24

Journal of Physics: Conference Series **206** (2010) 012033 doi:10.1088/1742-6596/206/1/012033

Introduction of a Minkowski space structure for a deeper insight in Euclidean issues

Jean-Marie Becker[1,2]

[1]Ecole Supérieure de Chimie Physique Electronique de Lyon, 69616 Lyon, France.

[2]Laboratoire Hubert Curien, CNRS UMR 5516, Université de Lyon, Université Jean-Monnet, 42000 Saint-Etienne.

E-mail : becker@cpe.fr

Abstract. $n \times n$ symmetrical positive definite matrices are present in many applications. In the $n=2$ case, these matrices constitute a 3D-space, denoted by S_2^+. The determinant of $A \in S_2^+$ is a natural quadratic form, giving S_2^+ a Minkowski space structure. This paper shows how this theory quite naturally provides a graphical interpretation to formulas that have been defined elsewhere. An original application to discrete curves is given.

1. Introduction

The norm N of a Euclidean space is defined by $\|x\|^2 = N(x)^2 = \sum_{k=1}^{n} x_k^2$ with respect to a certain basis $e_1,\cdots e_n$ where $x = \sum_{k=1}^{n} x_k e_k$ with associated dot product $b(x,y) = \sum_{k=1}^{n} x_k y_k$. A Minkowski space constitutes an extension of the Euclidean space structure. It is defined by the square of its "pseudo-norm" : $q(x) = q(x_0, x_1,\cdots x_n) = -x_0^2 + (x_1^2 + \cdots + x_n^2) = -x_0^2 + \|r(x)\|^2$ in a convenient $(n+1)$-dimensional basis. Of course, $\sqrt{q(x)}$ only exists in the "cone" defined by $x_1^2 + \cdots + x_n^2 \geq x_0^2$. The associated dot product to $q(x)$ is $b(x,y) = -x_0 y_0 + \sum_{k=1}^{n} x_k y_k$. The hyperplane with equation $x_0 = 0$, and more generally all affine hyperplanes with equations $x_0 = const.$, have a natural Euclidean structure, by taking $\|r(x)\|$ for its norm instead of $\sqrt{q(x)}$. Minkowski spaces occur quite naturally in several cases (relativity, space of spheres [1], etc…). This paper demonstrates the interest of these structures, by giving a graphical interpretation to ananlytical formulas given in [2].

2. A graphical representation of S_2^+

To a matrix $A = \begin{pmatrix} a & b \\ b & c \end{pmatrix}$ of S_2^+, one can associate the Minkowski quadratic form (see [1]):

$$q(A) = -4\det(A) = 4(b^2 - ac) = -(a+c)^2 + (a-c)^2 + (2b)^2 = -x_0^2 + x_1^2 + x_2^2 \qquad (1)$$

Thus $\begin{pmatrix} a & b \\ b & c \end{pmatrix} = x_0\sigma_0 + x_1\sigma_1 + x_2\sigma_2$ with matrices $\sigma_0 = \frac{1}{2}\begin{pmatrix} 1 & 0 \\ 0 & 1 \end{pmatrix}$ $\sigma_1 = \frac{1}{2}\begin{pmatrix} 1 & 0 \\ 0 & -1 \end{pmatrix}$ $\sigma_2 = \frac{1}{2}\begin{pmatrix} 0 & 1 \\ 1 & 0 \end{pmatrix}$.

In this way, S_2^+ can be represented as the interior of a cone in a 3D space (see Fig. 1).

The restriction of the norm to the Euclidean affine plane $x_0 = 1$ is shown to be $\|r(A)\| = \sqrt{x_1^2 + x_2^2} = \sqrt{(a-c)^2 + (2b)^2} = \sqrt{1 - 4\det(A)} = \lambda_{max} - \lambda_{min}$ (eigenvalues difference) and the Euclidean part of the dot product is $b(A, A') = x_1x_1' + x_2x_2' = (a-c)(a'-c') + 4bb'$ $= \|r(A)\|.\|r(A')\|\cos 2\theta$ (see Fig. 1 for the visualization of angle doubling). All these formulas are given in [2] without geometrical insight.

Now, let us consider a digital plane curve, defined by a file of consecutive neighbour points (x_k, y_k). To each value k_0 of k, one can associate the "cloud" made of 2N+1 points of the file, by taking all points with indices $k_0 - N \le k \le k_0 + N$, with a fixed N, for example with $N=2$ as represented on Fig. 1. To such a cloud, one attach its variance (and covariance) matrix $V \in S_2^+$, and normalize it in such a way that its trace is fixed.

Imposing trace $x_0 = 1$, each variance matrix can be represented as a point inside a "unit disk", intersection of cone S_2^+ with the plane $x_0 = 1$. Thus, to the initial plane curve, is associated, through the associated (normalized) V matrices, an abstract curve situated in a 2D representation space ; moreover, the evolution of this curve could be helpful in multiscale contour analysis.

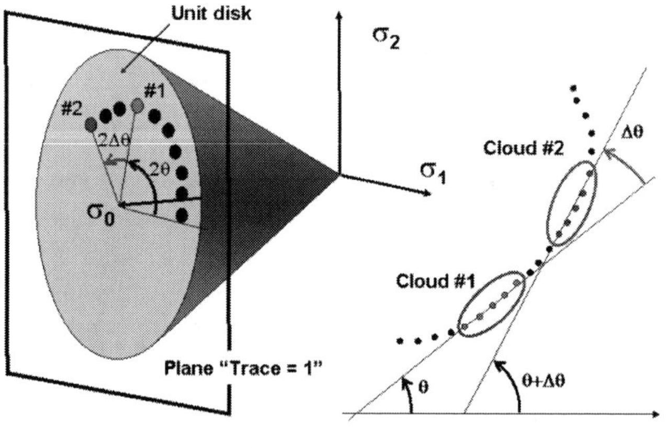

Figure 1: The "authorized" cone, sliced by plane $x_0 = 1$ with an application to digital curves.

References

[1] Berger M, Geometry I, Universitex, Springer, 2009 (chapter 20)..

[2] Anderson I.M.and Bezdek J.C. Curvature and tangential deflection of discrete arcs: a theory based on the commutator of scatter matrix pairs and its application to vertex detection in planar shape data. IEEE Transactions PAMI-6 27-40 (1984).

Geometric diffusion as a classifier*

Alan Van Nevel

1900 N Knox Rd,M/S 6302, Naval Air Warfare Center, Weapons Division, China Lake, CA 93555, USA

E-mail: alan.vannevel@navy.mil

Abstract. Often, the problem of automatic target recognition can be reduced down to two separate but related problems, feature extraction and classifier design. The best classifier only works as well as the input data provided to the system. In this presentation, we will outline a new approach to classification known as geometric diffusion as proposed by Coifman et al, and demonstrate the power of this new metric for classification of imagery.

1. Introduction

In this note, we present some results on object recognition which are very good and move away from traditional machine learning approaches. We will first discuss geometric diffusion and its use as a classifier. The example problem of interest, vehicle recognition in laser radar (ladar) imagery will be described, followed by a discussion of the results.

2. Geometric Diffusion

Geometric diffusion [1,2] can be thought of as a dimensionality reduction technique which preserves the intrinsic geometry of the data under study. The approach is similar to other dimensionality reduction techniques such as Laplacian Eigenmaps [3], Diffusion wavelets [4], and Spectral Clustering [5]. The approach in geometric diffusion is to develop a data dependent kernel which utilizes the spectral properties of the kernel to determine the underlying geometry of the data. This technique offers several advantages for use in dimensionality reduction, feature extraction and manifold applications. The main advantage is an extrinsic distance measure is used to recover the intrinsic data geometry, without the use of an *a priori* model for the data.

2.1. Basic Theory

The goal is to embed the data into a lower dimensional space, using a diffusion operator constructed from the data dependent heat kernel. The kernel **K,** needs to meet the following requirements: symmetric, positive semi-definite, and positivity preserving. A typical kernel that can be used to define distances between data points is

$$K(\boldsymbol{x}_i,\boldsymbol{x}_j)=\exp\left(-\frac{\left\|\boldsymbol{x}_i-\boldsymbol{x}_j\right\|^2}{2\varepsilon}\right) \tag{1}$$

* Cleared for public release, distribution unlimited

where $\boldsymbol{x}_i$, $\boldsymbol{x}_j$ represent the data under study. Other kernel functions can be defined as needed as long as they meet the conditions. The diffusion operator then is written as

$$A = \left[\frac{K(x_i, x_j)}{\sum_j K(x_i, x_j)} \right] = D^{-1}K \tag{2}$$

It is this operator A, that is critical for the classification task. Given a partially labeled data set X $=\{\boldsymbol{x}_i\}$, i = 1… N, where N is the number of data points, consider a label function $\chi_j(x)$ that equals 1 when x is a member of class j, and is zero otherwise. The procedure then is to smooth $\chi_j(x)$ using powers of A

$$\chi_i' = A^t \chi_i \tag{3}$$

where the power t is chosen typically by cross validation.

In this approach, not all of the data needs to be assigned a label. As long as each class has one exemplar provided, the smoothing operation will attempt to assign labels to each data point based on the underlying geometry of the data [6].

3. Ladar problem and results

Ladar sensors have the capability to capture 3-d information, which should provide for increased object recognition capabilities. Using synthetic ladar imagery, a five class recognition problem was developed, using 5 different military vehicles. Two different data sets were constructed, one where the depression angle varied from 5 to 52.5 degrees, at a fixed aspect, and the second had a fixed depression angle, and the aspect varied from 45 to 135 degrees. Each class had 20 examples. The imagery consisted of 32x32 pixel images. The images were lexicographically ordered by column, and the diffusion operator A constructed.

The performance was assessed as a function of the number of labelled exemplars provided per class, and averaged over 1000 runs, randomly selecting the images to be assigned a label. The remaining data was then classified by smoothing the label function with the diffusion operator. In each data set, as the number of labelled exemplars approached the maximum of 19 (equivalent to leave one out validation), the classification performance became 100%, and remained above 95% using as few as 7 exemplars. Using only one exemplar, performance was above 70% for the depression data

4. Conclusions

The geometric diffusion ideas provide a powerful new approach to clustering high dimensional data into a lower dimensional space while preserving the underlying intrinsic geometry. The approach permits improved classification performance for difficult image recognition tasks. The major drawback to this approach is the limited ability to add new data to the kernel. For limited recognition tasks, using small data sets, one can simply recompute the kernel, which quickly becomes unfeasible when dealing with very large high dimensional data.

References

[1] Coifman R and Lafon S 2006 *Appl Comp Harm Anal* **21** pp 5-30
[2] Lafon S "Diffusion Maps and Geometric Harmonics", Ph.D. Dissertation, Yale Univ, 2004
[3] Belkin M and Niyogi P 2003 *Neural Computation* **15**(6) pp. 1373-96
[4] Coifman R and Maggioni M 2006 "Diffusion Wavelets",*Appl Comm Harm Anal* **11**
[5] Chung F, "Spectral Graph Theory", CBMS-AMS, **92**, 1997
[6] Szlam A, Maggioni M and Coifman R 2008 *J.l of Machine Learning Research* **9** pp1711-38

2009 Euro-American Workshop on Information Optics IOP Publishing
Journal of Physics: Conference Series **206** (2010) 012035 doi:10.1088/1742-6596/206/1/012035

Robust visual hashing via ICA

Thierry Fournel[1], **Daniela Coltuc**[2]

[1] Université de Lyon, Université J. Monnet, Laboratoire Hubert Curien, UMR CNRS-UJM 5516, 18, rue du Pr Lauras, 42000, Saint-Etienne, France

[2] Faculty of Electronics, Telecommunication and Information Technology, University Politehnica of Bucharest, Bd. Iuliu Maniu nr. 1-3, Bucharest, Romania

E-mail: fournel@univ-st-etienne.fr

Abstract. Designed to maximize information transmission in the presence of noise, independent component analysis (ICA) could appear in certain circumstances as a statistics-based tool for robust visual hashing. Several ICA-based scenarios can attempt to reach this goal. A first one is here considered.

1. Motivations

We suggest here to investigate Independent Component Analysis (ICA) [1] for robust visual hashing. This approach may be relevant as ICA can efficiently capture features of data in various applications. As both the mixing matrix and Independent Components (ICs) are unknown in the ICA linear model, all ICs are assumed to have a unit variance therefore the same energy. This model provides a factorial representation of data having J statistically (quasi) independent components. Thus, some vectors could be formed by uniform quantization of the ICs, having an equal distribution and the property of distinction of visually different images [2].

Besides, designed to maximize information transmission in the presence of noise, ICA representations can be tolerant to variations giving visually similar images. For instance, such representations can be robust to expression changes in face recognition [3].

Let us build a uniform quantizer of the ICs from the ICA architecture 2 scheme in perspective to get J-dimensional intermediate hash values [4] in the context of face images.

2. A first scenario

The deviation with respect to the mean of a representative subset of face images, named the robot face, are first considered as decorrelated data (rather than raw face images themselves) for learning both weight matrix W (the pseudo-inverse of the mixing matrix) and ICs matrix S. By modeling the distribution of the independent components according to a Generalized Gaussian Distribution (GGD), q quantization intervals can be determined from matrix S such to have a uniform discrete distribution in $\{0, .., q-1\}$.

Given a face image x after substraction of the robot face, an intermediate binary hash value can be defined by concatenation of the Gray codes of the indices of the quantization intervals in which the ICs $s = W\,x$ are (Gray encoding allows a better robustness). Let us emphasize that a single unidimensional quantizer Q (defined at the learning phase) is here applied on each IC as ICs have all a unit variance. The hash values are decorrelated as face images are substracted to the robot face before computing ICs. The main advantage of the decorrelation is to provide a GGD distribution giving the possibility to design a quantizer such to have a uniform distribution for the components of the hash value and, further, due to components independency, to obtain uniformly distributed hash values.

Face Recognition DataBase from [5] was used for preliminary testing this approach. This database contains face images under a portrait format with considerable expression changes for any given individual (by asking the individuals to speak). In this experiment, the images of the different poses of the same individual are considered as similar images whereas the images representing different individuals are considered as different ones. The training set contains the images of 60 individuals (40 boys, 10 girls and 10 old persons), each in 5 different poses. The test set is formed from 40 of the individuals involved in the training set plus 10 new ones, each of them in 10 new poses. The face zone was cut from each image and resized to 42x28 pixels.

100 ICs and 8 bins have been considered for quantization (300 bits per hash value i.e. a dimensionality reduction rate of 3.2%). The parameters of the GGD of the ICs have been estimated to $\hat{\alpha} = 15.4$, $\hat{\beta} = 1.2$ from the training set. The histograms of the normalized Hamming distance between the binary hash vectors measured on the test set for the similars and non-similars, respectively, intersect. The Hamming distance present a short bias (3%) when comparing non-similars each other. However for similars, its mean value is clearly high (11.7%). This can be explained by residual synchronization errors between the images of the different poses of individuals and by the number of ICs to adjust.

3. Conclusion

Some constraints on data (as only recording neutral expressions) could be assumed and / or refinements inserted in order to improve the quality of intermediate hash values which can be obtained by ICA. As ICA tends to give words uniformly distributed in a J-dimensional space, the final step can be ideally result from tiling the space with spheres having the required error correction rate as radius. By this way, the final hash value is given by the centre of the sphere in which the intermediate hash value is.

For authentication applications, key-dependent hashing can be achieved by randomizing hash components from a secret seed. A pseudo-random reordering can be performed on the ICs then on the bits of the indices at the output of the quantifier.

References

[1] Hyvarinen A and Oja E 2000 *Neural Networks* **13:4-5** 411-430
[2] Skrepth C J and Uhl A 2003 *Pattern recognition and image analysis* **2652** (Springer Berlin) 986-993
[3] Bartlett M S, Movellan J R and Sejnowski T J 2002 *IEEE Trans. on Neural Networks* **13:6** 1450-1464
[4] Venkatesan R and Mihcak K 2002 *Security and Privacy in Digital Rights Management* **2320** (Springer Berlin/Heidelberg) 13-21
[5] *Face Recognition DataBase* from University of Essex (UK) at http://cswww.essex.ac.uk/mv/allfaces/index.html

Secure embedded verification of print signatures

Abdourhamane Idrissa, Thierry Fournel, Alain Aubert

Université de Lyon, Université J. Monnet, Laboratoire Hubert Curien, UMR CNRS-UJM 5516, 18, rue du Pr Lauras, 42000, Saint-Etienne, France

E-mail: `fournel@univ-st-etienne.fr`

Abstract. This work aims to embed an object authentication process in a secure electronic architecture for distant verification via imaging. Firstly, the extraction of *print signatures* [1] is revisited for an implementation in a logical hardware system. Secondly, a protocol scheme to authenticate the verification device and to ensure its integrity is suggested.

1. Motivations

To verify object authenticity through features with an autonomous device i.e. without communicating with authority for decision, the device needs the computing ressources required for feature extraction, feature matching, data checking and device integrity testing.

Zhu et al. described in [1] a local process involving the first three functionalities. This system verifies the authenticity of laser-printed documents (like tickets) through both a cryptographic digital signature [2] and a "print signature" resulting from the non-repeatable shape obtained when laser-printing [3] a given pattern.

We reconsider here the shape extraction in the framework of hardware implementation and integer computations in an embedded device, then we suggest a variant working in the Discrete Cosine Transformation (DCT) domain. Their respective performances are compared on an experimental set in section 2.

Another crucial point, specific to autonomous verification, is to check the consistency of the device. For this purpose, we introduce two major blocks into a special architecture: one block dedicated to device certification and another one to device integrity check (Figure 1). Afterwards, we express the device authentication protocol in two parts: identifying the verification device (via the first block communicating with the authentication server) and checking its integrity.

2. Feature extraction

The feature extraction performed in [1] to exploit a set of dots as print signature consists in computing a set of normalized radii computed from the centroid of each dot. We can release the normalization (by the average radius of the shape) by fixing the magnification of the recording system in order to tract integer values as feature descriptor (a number of significative digits). To relax the constraint of a star shape (as dot), we suggest a variant only based on the edge (without using the centroid). The edge C of a given dot (a list of integers coordinates) is divided in four sectors $\mathbf{C}^j = (\mathbf{C}^j_x + i \cdot \mathbf{C}^j_y), j \in \{1, 2, 3, 4\}$, each described by a vector of N digitalized DCT coefficients: $DCT_i(\mathbf{C}^j), i \in \{1, \ldots, N\}$. The shape descriptor is formed by concatenation of the four DCT coefficient vectors associated to the four sectors ordered in the sense of clockwise.

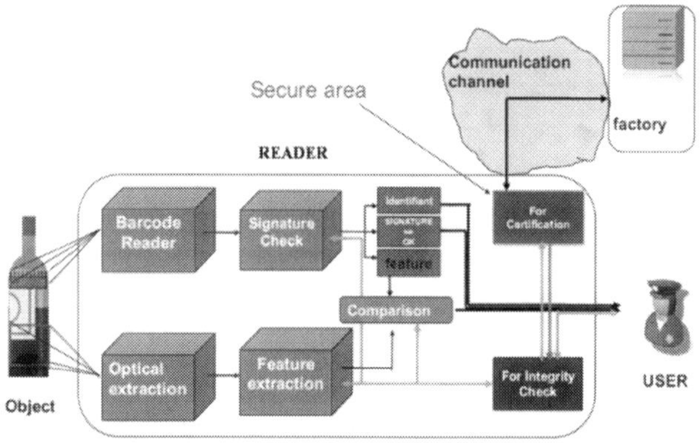

Figure 1. Blocks of the electronic architecture suggested for secure embedded verification of objects.

To compare the trends of the two types of features, the integer version of the radius-based features and the integer DCT-based ones, a set of images of $300\mu m$ laser-printed dots was recorded in experimental conditions close to the ones in [1]. The pairwise comparisons of features extracted from images of the same dot, different dots, respectively, were performed via the Euclidian distance (Figure 2). The trends are closely similar for both types of features, with identical ambiguous cases due to some locally strong image variations at the edge of the shape.

3. Integrated architecture for verification and protocol scheme

To perform a secure distant control, the ability of the embedded authentication device (named hereafter the Reader) to make cryptographic verification (with valid public keys), to extract and match features with integrity must be checked. In the protocol we suggest that the Factory plays the role of authority and is in charge to check the Reader ability. In the first step (connection phase), a query filed by Reader identification number ID_R and encrypted data $ENCRYPT_{K_R}(ID_R, N_R)$ are sent by the Reader to the Factory in order to establish a secure communication by generating a nonce N_R and by using secret symmetric key K_R. Once the connection established, the Reader integrity is checked by the Factory according to a zero-knowledge scheme [2, 4] as follows:

$$\left.\begin{array}{llll}
1. & Factory \rightarrow & Reader: & ENCRYPT_{K_{\texttt{session}}}(img, f, index, \texttt{SIGN}(f)) \\
2. & & Reader: & f' := \texttt{FEATURE}(img) \\
 & & & b_0 := \texttt{MATCH}(f, f') \in \{0,1\} \\
 & & & b_1 := \texttt{CHECKSIGN}(f, index, \texttt{SIGN}(f)) \in \{0,1\} \\
 & Reader \rightarrow & Factory: & ENCRYPT_{K_{\texttt{session}}}(\texttt{SIGN}(f) \,||\, b_1 \,||\, b_0) \\
3. & & Factory: & \textbf{If } (\texttt{SIGN}(f) \,||\, b_1 \,||\, b_0) \textbf{ Is Valid Then } \texttt{replay 1.} \\
 & & & \textbf{Else } \texttt{Stop communication.}
\end{array}\right\} \texttt{n times}$$

where firstly img, f and f' denote a test image, its registered feature descriptor and a re-extracted one, respectively, secondly modules $ENCRYPT$, $SIGN$, $CHECKSIGN$,

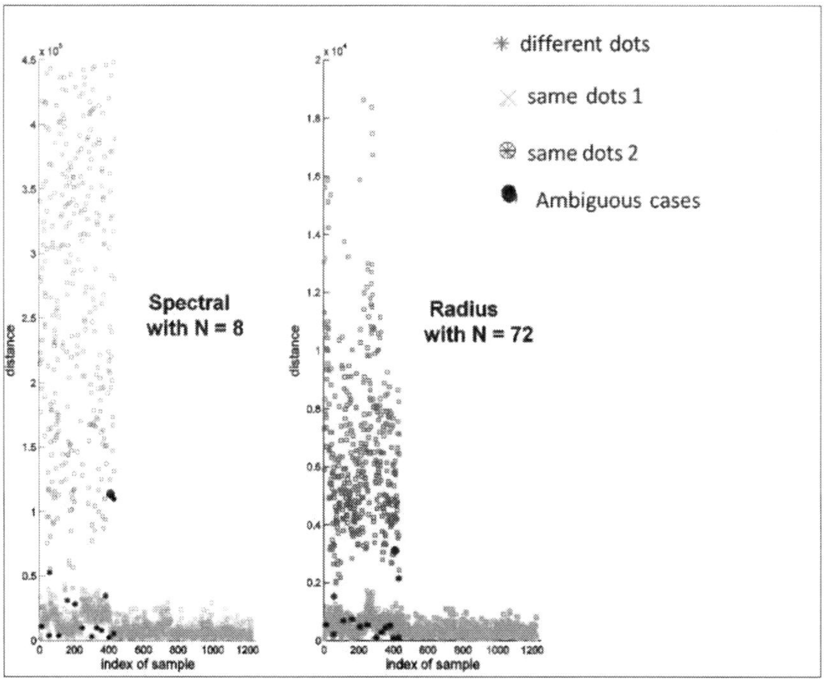

Figure 2. Pairwise comparisons of features via the Euclidian distance for the integer DCT-based features (N=8 digitalized DCT coefficients per sector)(left) and the integer radius-based ones (N=72 radii per edge) (right).

$FEATURE$ and $MATCH$ are dedicated to encryption, digital signature, digital signature verification, feature extraction and matching, respectively.

4. Conclusion and perspectives

In comparison with the feature extraction method suggested in [1], the spectral version here presented (via DCT) has similar results but it extends to a broader set of print signatures namely non star shapes. The Reader-Factory protocol suggested to verify device authenticity/integrity has now to be proved.

Acknowledgements

This work supported by Region Rhne-Alpes has been developped in the frame of the Cluster ISLE / Project SISS / Operation TraSecure (2006-2009). The authors would also thank Signoptic Technologies for the fruitful exchanges they had with it on the topic.

References

[1] Baoshi Zhu, Jiankang Wu and Mohan S. Kankanhalli. Print Signatures for Document Authentication. *In Proc.10th ACM Conference on Computer and Communications in Security,(CCS'2003)*,pages 145-153. ACM Press, 2003.

[2] A. J. Menezes, P. C. van Oorschot, and S. A. Vanstone. *Handbook of Applied Cryptography*. CRC Press, 1997.

[3] S. J. Bigelow and E. Kuaimoku. *Easy Laser Printer Maintenance and Repair*. Windcrest/McGraw-Hill, Blue Ridge Summit, PA, 1994.

[4] Jean-Jacques Quisquater, Louis C. Guillou, Thomas A. Berson. How to Explain Zero-Knowledge Protocols to Your Children. *Advances in Cryptology - CRYPTO '89: Proceedings*, v.435, pages 628-631, 1990.

2009 Euro-American Workshop on Information Optics IOP Publishing
Journal of Physics: Conference Series **206** (2010) 012037 doi:10.1088/1742-6596/206/1/012037

Watermarking Scar as an Ultimate copy protection

Sofiane Braci, Abdelkader Miraoui, Claude Delpha and Rémy Boyer

University Paris-Sud 11, Laboratory of Signals and Systems, Supélec, 3 rue Joliot-Curie, 91192 Gif-sur-Yvette cedex, France.

E-mail: {sofiane.braci, Abdelkader.Miraoui, claude.delpha, remy.boyer}@lss.supelec.fr

Abstract. Strong and powerful attacks are able to make the watermark extraction impossible even if there is some watermarking. In this work, we call this residual information the scar and we use it to prove the existence of an attacked watermark. We evaluate the Scar by the mutual information between the embedded watermark and the attacked copies. It is well known that the mutual information between two random variables gives an evaluation of the shared information between the two variables. Thus, the attack can compromise the correct decoding by removing partialy the watermarking information. However, it is possible to make the watermarking existence proof possible when the mutual information between the embedded message and the attacked copies is above a certain threshold. We propose a practical way to use the scar by measuring the correlation between the attacked watermark and the original one.

1. Introduction

We can class the watermarking attacks into two important kinds, the first one, includes the attacks based on the estimation of the watermarking secrecy, in order to produce an illegal copy as proposed in the paper [1]. The second kind of attacks try to remove the watermarking without any knowledge about it (secret parameter, localizations,...etc.), as the Temporal Frame averaging (TFA) (for example on video watermarking, see paper [2]). It is well known that it is impossible to reach the perfect watermarking security. In the other hand, a perfect removal attacks which perfectly "washes" the watermarked copies and produce hacked copies without any watermark is also impossible. Generally, it is due to perceptual and computational constraints. The aim of this work is to use the residual information of the attacked (partially removed) embedded mark as a "Scar" to prove the previous existence of watermarking.

2. Using the watermarking scar to enhance the watermarking systems performance

The scar is a strong way to prove the watermarking existence and permits to proceed to an additional processing for the watermarking recovery. Thus, the watermarking system is more effective if the attacker have to spend a lot of resources and effort to remove all residual information (the scar). If we choose a watermarking system which leaves many embedding information after attacks, we have more chance to retrieve our watermark information and make the copies protection stronger. The watermarking scar can be measured thanks to the mutual information between the embedded watermark and the marked/attacked copies. We compute this mutual information for two different systems SCS [3] and QIM [4] as shown in Fig.1(a).

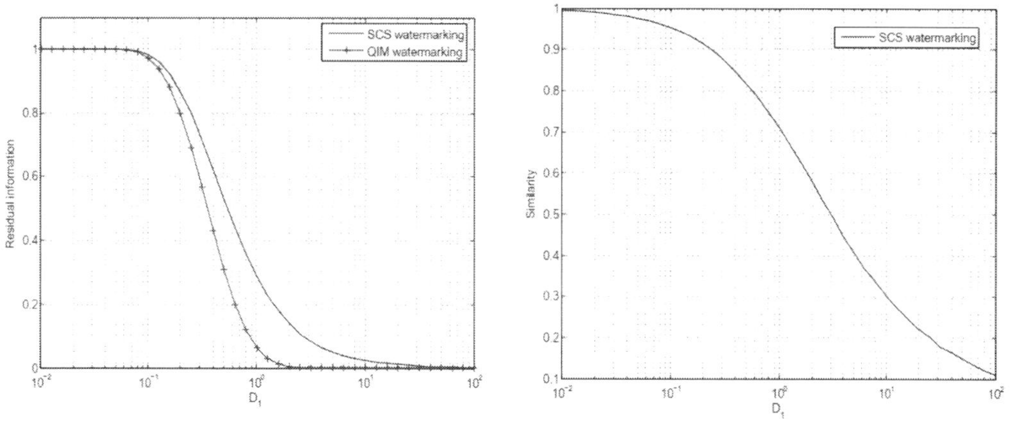

(a) (b)

Figure 1. (a) The residual information (scar) of the embedded watermark after and AWGN attack as funtion of the additive noise strength D_1 (b) The similarity -given by the normalized correlation- between the watermarked/attacked signal and the watermarknig signal as funtion of the additive noise strength D_1.

As expected the scar strength fails when the noise strength increases. However, we note that the scar is slightly more important for SCS watermarking than the QIM watermarking. It is due the fact that the SCS embedding is more robust and resistant than the QIM embedding. Thus, the embedding information has more chances to survive against an AWGN attack with the SCS than the QIM. In Fig.1(b), we have computed the correlation between the attacked watermark represented by the vector $\dot{\mathbf{w}}$ and the embedded watermark $\mathbf{w}$. From Fig. 1(b), we note that the correlation between the attacked QIM watermark and the original one fails when the additive noise strength increases. In addition, the correlation curve decreases and follow the same behavior of the residual information (Scar) given in Fig. 1(a). The correlation is a practical way to use the watermarking scar after a malicious attack. It is possible to fixe some threshold for the correlation to decide if there is enough evidence of the watermarking existence and proceed for example to a denoising processing to retrieve the watermark.

3. Conclusion

Through this work, we introduce a new notion : The watermarking Scar. We translate an intuitive idea present in many works on watermarking security. A definition of the Scar watermarking notion and the information theory tools for evaluation are given. The obtained results show that the watermarking scar notion can become a criterion for watermarking system. They prove that it is possible to retrieve the watermarking even if the attacker proceeds to a strong attacks which makes the watermarking decoding impossible.

References

[1] F. Cayre, C. Fontaine, and T. Furon. Watermarking security: Theory and practice. *IEEE Trans. on Signal Processing*, 53:3976–3987, 2005.

[2] R. Caldelli, A. Piva, M. Barni, and A. Carboni. Effectiveness of st-dm watermarking against intra-video collusion. In *Digital Watermarking. Lecture Notes in Computer Science*, pages 158–170. Springer-Verlag, 2005.

[3] J. J. Eggers, R. Bauml, R. Tzchoppe, and B. Girod. Scalar costa scheme for information embedding. *IEEE Trans. on Signal Processing*, 51:1003–1019, 2003.

[4] B. Chen and G. W. Wornell. Quantization index modulation: a class of provably good methods for digital watermarking and information embedding. *IEEE Trans. on Information Theory*, 47:1423–1443, 2001.

Overview of FTV (Free-viewpoint Television)

Masayuki Tanimoto

Graduate School of Engineering, Nagoya University, Furo-cho, Chikusa-ku, Nagoya, 464-8603 Japan

E-mail: tanimoto@nuee.nagoya-u.ac.jp

Abstract. We have developed a new type of television named FTV (Free-viewpoint TV). FTV is an innovative visual media that enables us to view a 3D scene by freely changing our viewpoints. FTV is based on the ray-space method that represents one ray in real space with one point in the ray-space. We have also developed new type of ray capture and display technologies such as a 360-degree mirror-scan ray capturing system and a 360 degree ray-reproducing display. The international standardization of FTV has been conducted in MPEG.

1. Introduction

We have developed a new type of television named FTV (Free viewpoint TV) [1], [2]. FTV is an innovative visual media that enables us to view a 3D scene by freely changing our viewpoints as if we were there. FTV will bring an epochal change in the history of visual media.
The standardization activity of FTV has been conducted at MPEG since 2001. The first phase of FTV was Multi-view Video Coding [3] and the second phase is 3D Video [4].

2. FTV System

The FTV system captures a 3D scene by multiple cameras on the sender side. A "100-camera system" was developed to capture large space. The captured images contain the misalignment and luminance differences of the cameras. They are corrected. The corrected images are compressed and transmitted.

On the receiver side, the images are reconstructed by decoding. The ray-space is constructed by arranging the reconstructed images and interpolating them. Free-viewpoint images are generated by cutting the ray-space. FTV players are realized on a PC or a portable player as shown in Fig.1.

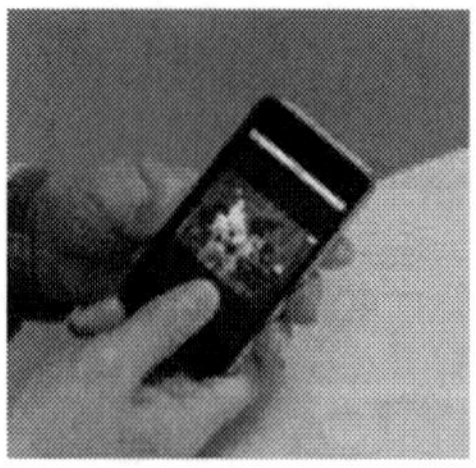

Figure 1. FTV on a PC (left) and FTV on a portable player (right).

3. Ray-reproducting FTV

We are developing ray-reproducing FTV. Rays are captured by a 360-degree mirror scan ray-capturing system [5]. It acquires a dense ray-space in real time without interpolation. In this capturing system, a high-speed camera and a scanning optical system are used instead of multiple cameras. Rays are reproduced by a 360-degree, ray-producing display SeeLINDER [6].

4. International Standardization of FTV

Fig.2 shows the history of FTV standardization at MPEG. The standardization activity of FTV has been conducted since 2001. The first activity was MVC (Multi-view Video Coding) that is the standardization of the coding part of FTV. The second activity is related to the view generation of FTV. It is 3DV (3D Video) and targets serving for a variety of 3D displays. MVC was completed in May 2009.

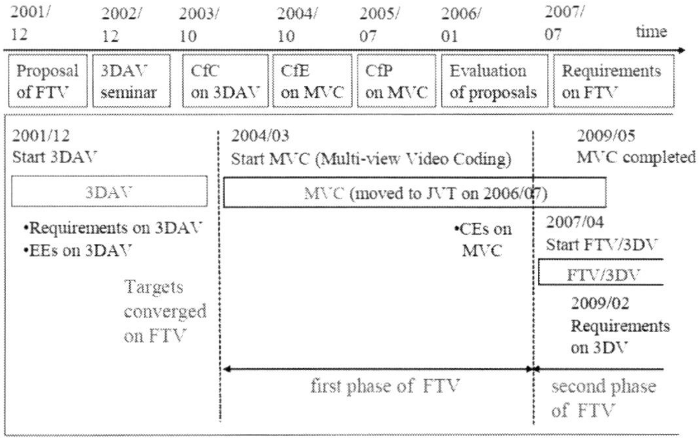

Figure 2. History of FTV standardization in MPEG.

5. Conclusion

FTV is the ultimate 3DTV, a natural interface between human and environment, and an innovative tool to create new types of content and art. FTV will find many applications in the fields of broadcast, communication, game, entertainment, advertising, exhibition, education, medicine, art, archives, security, surveillance, and so on.

This research is partially supported by Strategic Information and Communications R&D Promotion Programme (SCOPE) 093106002 of the Ministry of Internal Affairs and Communications, and National Institute of Information and Communica-tions Technology, Japan (NICT).

References

[1] Masayuki Tanimoto, " Overview of Free Viewpoint Television, " Signal Processing: Image Communication, vol. 21, no. 6, pp. 454-461, July 2006

[2] Masayuki Tanimoto; "FTV (Free viewpoint TV) and Creation of Ray-Based Image Engineering", ECTI Transaction on Electrical Engineering, Electronics and Communications, Vol. 6, No. 1, pp.3-14 (February 2008).□Invited Paper□

[3] "Introduction to Multi-view Video Coding", ISO/IEC JTC 1/SC 29/WG11, N7328, July 2005.http://www.chiariglione.org/mpeg/technologies/mp-mv/index.htm

[4] "Introduction to 3D Video", ISO/IEC JTC1/SC29/WG11 N9784, May 2008. http://www.chiariglione.org/mpeg/technologies/mp3d/index.htm

[5] Keisuke Manoh, Tomohiro Yendo, Toshiaki Fujii, and Masayuki Tanimoto, "Ray-Space Acquisition System of All-Around Convergent Views using a Rotation Mirror", Proc. of SPIE, Vol. 6778, pp.67780C-1-8, September 2007.

[6] T. Endo, Y. Kajiki, T. Honda and M. Sato, "Cylindrical 3-D Video Display Observable from All Directions", Proc. of Pacific Graphics 2000, pp. 300-306, October 2000.

AUTHOR INDEX

AUTHOR INDEX

Institute of Physics
Dirac House, Temple Back
Bristol BS1 6BE UK

CURRAN ASSOCIATES INC
proceedings.com